ALS
Advances in Life Sciences

Species Conservation: A Population-Biological Approach

Edited by
A. Seitz
V. Loeschcke

Birkhäuser Verlag
Basel · Boston · Berlin

Editors' addresses:

Prof. Alfred Seitz
Institut für Zoologie
der Universität Mainz
Saarstrasse 21
D-6500 Mainz
Germany

Prof. Volker Loeschcke
Institute of Ecology
and Genetics
University of Aarhus
Ny Munkegade
DK-8000 Aarhus C
Denmark

The use of registered names, trademarks, etc. in this publication does not imply, even in the absence of a specific statement, that such names are exempt from the relevant protective laws and regulations and therefore free for general use.

Deutsche Bibliothek Cataloging-in-Publication Data
Species conservation: a population-biological approach / ed. by A. Seitz; V. Loeschcke. — Printed from the author's camera-ready ms. — Basel; Boston; Berlin: Birkhäuser, 1991
(Advances in life sciences)
ISBN 3-7643-2493-7 (Basel ...)
ISBN 0-8176-2493-7 (Boston)
NE: Seitz, Alfred [Hrsg.]

© 1991 Birkhäuser Verlag
        P.O. Box 133
        4010 Basel
        Switzerland

Printed from the authors' camera-ready manuscripts
on acid-free paper in Germany
ISBN 3-7643-2493-7
ISBN 0-8176-2493-7

Contents

Preface

The large expansion of the human population in the last few hundred years has caused drastic changes in the environment with impact on the biota of the whole earth. Humans have become the dominating species, and consequently the survival of thousands of other species is threatened. Extinction rates have increased considerably and biological diversity has decreased accordingly, as the loss of species diversity was not any more balanced by speciation. (From the estimated 8 - 30 Million species existing on earth, only a minor fraction of about 1,5 million is yet described. If the fragmentation and destruction of habitats continues, a huge number of species will disappear before we ever get the chance to learn about their existence.)

The loss of species diversity has recently drawn increased attention from scientists, politicians and also a portion of the general public. In this book population biologists have contributed from their field of experience on issues relevant to the preservation of species diversity. Though species extinction is a global problem, it was focussed mainly on the European situation that has not received the same attention as conservation problems in tropical regions.

The source of this book is a symposium on "Populationsbiologische Aspekte des Artenschutzes" held in October 1989 at the Akademie der Wissenschaften und der Literatur in Mainz, FRG. The symposuim was funded by the German Minister of the Environment, Natural Protection and Reactor Safety (Bundesminsterium für Umwelt, Naturschutz und Reaktorsicherheit). The audience consisted of scientists interested in conservation biology as well as many people in one way or another active in conservation practice. The manuscripts of the contributions were reviewed by specialists in the field.

We are grateful to the German Minister of the Environment, Natural Protection and Reactor Safety for giving the necessary financial support, the Academy of Science and Literature in Mainz for help in the local organization and for providing the lecture hall and other facilities, the participants for many stimulating discussions, the simultaneous interpreters for doing an excellent job during the symposium and the reviewers for their critical comments on the manuscripts. Mrs. M. Becker and Dr. M. Klein did the laborous job checking the manuscripts, improving figures and arranging the layout of this volume.

Mainz and Armidale, January 1991   Alfred Seitz, Volker Loeschcke

Species Conservation: A Population-Biological Approach
A. SEITZ & V. LOESCHCKE (eds.) © 1991 Birkhäuser Verlag, Basel

Introductory Remarks: Population Biology, the Scientific Interface to Species Conservation

A. Seitz, Institut für Zoologie, AG Populationsbiologie, Johannes Gutenberg-Universität Mainz, Saarstr. 21, D-6500 Mainz, Germany

Abstract

Species conservation is one of the most challenging problems of present biological research. The scientific discipline "population biology" covers — with its methods and theoretical background — a large proportion of the tasks which are necessary to make proposals for practical conservation projects.

Examples are shown which demonstrate the potentials of population biological analyses on different levels of integration and with different experimental or analytical approaches. The plasticity of many population properties and the interrelations between ecological and genetical causes and consequences make it difficult in many cases to recognize key factors for the threatening of populations or species.

The stochasticity of the environment and the high number of interacting abiotic and biotic factors are the central problems from an ecological point of view which make long-term predictions impossible. More difficulties arise, if genetic properties are included. Natural and man-made changes of the environment shift the genetic structure of populations and cause a loss of genetic polymorphism. This results in a reduced ability of the populations to adapt to changing environments.

New analytical tools, primarily developed for forensic studies and population analyses, e.g. DNA fingerprinting, have opened up new ways to study gene flow on a very high level of accuracy. These methods can also be used for practical cases of species protection, e.g. paternity testing during breeding experiments, release of bred animals, control of resource populations or forensics.

Introduction

The continuous loss of animal and plant species and the decline of the population size of many more species is a global problem. At no time during the history of the earth was it more severe than during the time in which we are living. Actually, half of the vertebrates and one third of plant species are considered to be endangered. In many cases it is difficult to prove that a certain species has completely disappeared. As DIAMOND (1990) has pointed out, many more species have become extinct than we expected, because "most species live in tropical areas where few zoologists reside, and where extinctions are thus most likely overlooked". This development can not be tolerated. There are a number of reasons for the decline of species, and a number of precautions to avert this process have been, more or less half-heartedly, attempted.

I do not attempt to justify the worldwide campaign for species protection from a scientific point of view. It is difficult and probably impossible in many cases to assess the ecological consequences of the loss of rare and endangered species. I rather assume that the conservation of unique species that took millions of years to evolve has to be quite as important as the maintenance of cultural monuments for which a large amount of money is spent. Instead of arguing in favor of species protection, I will focus on a few topics in the field of population biology which could prove useful in solving problems of species conservation.

Two main causes are generally accepted as the reasons for the decline of species and their extinction:

1) Direct damage through environmental chemicals (xenobiotics) which are emitted in the course of industrial production, voluntarily for crop protection or by accident. This has led to a reglementation of the production and release of numerous chemicals by law in many countries to protect man and nature. The concentration of these chemicals, however, is often so low in nature that the direct toxicity *in situ* can hardly be demonstrated.

2) The loss of habitats in the course of changes of land use for agriculture, transport systems and growing cities.

A third point is often overlooked. This happens because even scientists who deal with this matter, have problems. It is the structural change of the environment and in the course of that, the change in the ways and amount of interactions within and between animal and plant populations. Such structural changes may be much more important and harmful than the quantitative changes of the environment.

The study of those interactions is the field of population biological research. Population biology is a comparably young scientific discipline. It was born out of the knowledge that populations are the units of evolution, where ecological and genetic processes take place that are necessary for the permanent coexistence of species, communities and ecosystems. The unification of evolutionary biology, population ecology and population genetics into the discipline "population biology" constitutes an enormous challenge for biologists, because in many cases it requires new ways of thinking, new methods and new scientific approaches. One central dogma of population biology is that the properties of species can only be understood correctly if one accepts the fact that above the level of individuals which are mainly the focus of physiological research projects, there are new levels of integration, the populations, which are essential for the survival of the species. The populations, of course, are characterized by the properties of the individuals, however they receive new ones due to this integration, e.g. age structure, genetic constitution, distribution in space and time.

Population biology has a number of tools which make it easier to deal with these new properties. They enable the study of populations in the field, and also in the laboratory by means of mathematical models and computer programs. In fact, theory plays a central role in the study of population behavior, because most real populations are not very suitable for experimental treatment.

The problem of population dynamics in space and time

We know from the study of natural and experimental populations that they are no static entities in space and time. The reasons for this are found in the variability of environmental factors and in processes which are inherent to populations, e.g. time lags which provoke regular or irregular oscillations of population density and age structure (SEITZ 1973, HALBACH 1979). Even in a very predictable environment, e.g. the pelagic area of lakes,

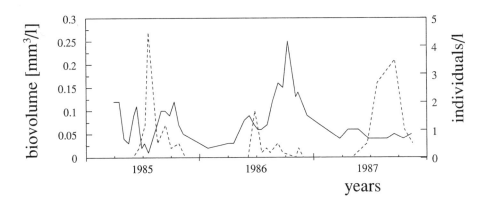

Fig. 1:   Population dynamics of zooplankton and phytoplankton in Lake Laacher See (Data from Eckartz-Nolden, 1988). Solid line= biovolume of algae which are edible for zooplankton and dashed line= number of *Daphnia*. Even in rather deterministic systems such as the pelagic areas of lakes, yearly cycles differ with respect to amplitude and phase.

large differences have been observed between years (fig. 1). Models are helpful in recognizing critical properties of ecological systems with respect to structure and function.

It seems, however, that the results of models are often overrated. Current simulation studies in our working group show that a large group of ecological systems can have a "quasi-chaotic" behavior which makes long-term predictions impossible. A useful approach may be the prediction of probabilities for the occurrence of certain system states (fig. 2).

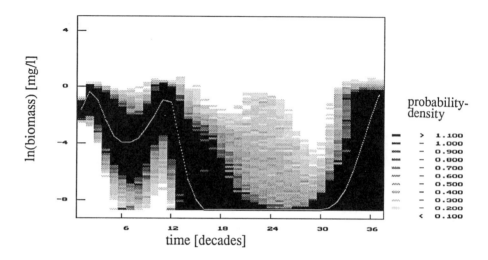

Fig. 2:   Probability density of system states of a numerically simulated zooplankton population. The model consists of five biotic compartments (two zooplankton, two phytoplankton, one fish compartment) and eight abiotic compartments (nitrogen and phosphorus). The model is driven by the seasonal fluctuations of solar radiation and temperature. The tremendous variation of system states are caused by just 10% random fluctuation of input variables. This system behavior demonstrates the problems which are associated with the prediction of "natural" population dynamics. From POETHKE et al. 1991.

From these reasons, serious problems arise for species conservation. We have to decide whether a low population density is caused by natural factors or by man. That requires a careful study of the population structure and dynamics. Long-term population cycles impose special problems. It is possible that an observed low population density is only the outcome of a natural cycle which had a minimum during the time of the study. On the other hand a population optimum can mask the damage already affecting the population and lead to an overrating of the endurance of populations. The consequence may be an unexpected rapid crash of the population density.

The problems of population structure

Without a sound knowledge of the ecological structure of populations and at least a rough idea about the driving forces of the population dynamics, it is impossible to assess the status of a species. There are good reasons to believe that most species do not exist in stable populations, but are subdivided into a network of interacting, locally instable sub-populations. These so-called "metapopulations" constitute special problems and potentials for species conservation. They have unique ecological and genetical properties. By example of populations of the tephritid fly *Urophora cardui* these properties will be discussed shortly.

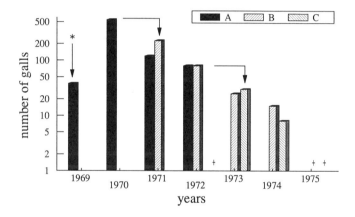

Fig. 3:   Development of experimental colonies of the tephritid fly *Urophora cardui* in a pasture in the Swiss Jura. Data from ZWÖLFER (1979). Colony A was founded in 1969 by the release of 120 flies (sex ratio 1:1). Subcolonies B and C were founded in 1971 and 1973, resp., by migrants from the existing colonies. The new subcolonies were founded within a distance of 40 - 50 m. Colony A was eradicated by Chalcoid parasites of the genus *Eurytoma*. Colonies B and C were finally destroyed by a farmer.

Urophora cardui oviposits into the stems of the creeping thistle *Cirsium arvense* and the developing larvae induce the formation of a gall from which they feed. The population density can therefore be estimated quite easily by counting the galls on the plants. Because *Urophora cardui* was expected to be a potential agent for the biological control of thistles, its population biology has been studied quite carefully. Because the fly uses only one host plant for oviposition, the populations are restricted to areas where this plant occurs. This results in a patchy arrangement of relatively small subpopulations. Observations of natural and experimental populations (fig. 3) proved that the subpopulations are unstable and become extinct through the action of parasitic Hymenoptera. Usually, however, *Urophora* founds a new subpopulation before the old one crashes. The continuous process of extinction and recolonization is only possible if the colonizaton rate is high enough and this is only possible if the density of possible habitats reaches a certain level.

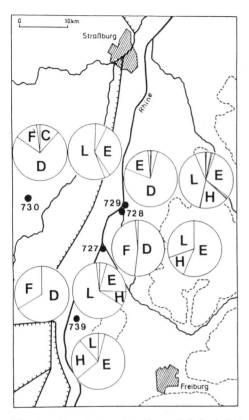

The example of *Urophora cardui* is a model of space-structured populations. The partial separation of subpopulations leads to a spreading of risk and reduces the probability of the whole population becoming extinct. As we know from theory, subdivision of populations results in a change of phenotypic and genotypic variability within the local subpopulations and between the subpopulations. Depending on the mode of colony formation, repeated extinction and recolonization can either decrease or increase the genetic differentiation of colonies (MCCAULEY 1991). With respect to the above example of *Urophora* populations, theory coincides with the results of the field studies. The small local subpopulations show pronounced differences at several enzyme loci (Fig. 4). Our increasing knowledge about the genetic structure of populations since the development of gel electrophoretic techniques (SMITHIES 1955, SHAW & PRASARD 1970) and their broad application in the last twenty years has led to new insights into the dynamics of gene flow between populations.

Fig. 4: Gene frequencies of *acon1* (left circles) and *got1* (right circles) of *Urophora cardui* populations in the upper Rhine valley. The letters designate representative electromorphs. Data for *acon1* are not available for station 739. (From SEITZ & KOMMA 1984)

The flow of individuals and genes between the subpopulations of a metapopulation is essential for its properties. Changes in the environment are therefore expected to have a serious impact on the dynamic stability of such systems. Such changes could be altered land use, construction of roads, or partial destruction of habitats or essential resources. Because different members in a community have different reactions to these changes, not only single populations are affected, but also the interactions between the populations and as a consequence the balance between the species.

Reduced gene flow leads to an isolation of the subpopulations and breaks the structure of a metapopulation. The result is a reduced genetic variability caused by inbreeding and genetic drift. This has been shown for populations of the common frog, *Rana temporaria* (tab. 1). In this study, the less isolated populations had more alleles and were more heterozygous than the more isolated ones. The barriers were in this case roads, railway tracks and motorways which blocked the migration effectively. Radio tracking of frogs during the off-pond migration after spawning showed that they avoid the embankments of roads and railways (DOEPNER 1990). The construction of "toad tunnels" is therefore only a solution if the roads are on the same level as the surrounding area.

Reduced ability to migrate has not only genetical consequences. As pointed out in the case of metapopulations, there must be a balance between colonization and extinction rate of subpopulations to achieve a dynamically stable system. Besides these medium to long-term effects, migration is a key factor for the survival of all species that perform regular migrations, e.g. spawning migrations of amphibians or changes of the habitat of a number of lepidopterans during their life cycle . Blocking of passages or destruction of secondary habitats may extinguish such populations within just one generation.

Tab. 1: Genetic variability, average heterozygosity and number of alleles of the common frog *Rana temporaria* in the Saar-Palatinate lowland (FRG) (REH & SEITZ 1990). Averages and standard deviation are given. The classification of the degree of isolation is based on the number of roads, railway tracks and motorways.

Degree of isolation	Number of populations	Population size	Degree of heterozygosity [‰]	Total number of alleles of 24 investigated loci	Monomorphic loci [%]
strong	6	55.8 56.9	7.05 5.82	46.17 10.44	38.33 10.52
medium	10	35.3 15.8	20.34 6.45	53.50 10.16	31.30 13.06
weak	4	77.5 47.6	28.30 7.50	54.25 5.56	32.50 14.71

The problem of genetic variability

It is generally accepted that loss of genetic information is an important burden for species and affects the prospect of their continuing existence. This seems to be true for a large number of species of probably all taxa. The assumption that a considerable amount of genetic variability is necessary, is based on the correlations between the abundance of genotypes and environmental factors. A typical example of a correlation between one environmental factor and the frequency of a genetic trait is shown in figure 5. From such findings one can conclude that a heterogeneous environment favors the maintenance of genetic variability. The trend may indeed be that generalists have a higher genetic variability than specialists (NEVO *et al.* 1984). Whether or not genetic variability influences short-term survival probabilities is obviously a different matter in each special case. Elephant seals (*Mirouga angustirostris*) seem to survive in spite of a lack of electrophoretically detectable polymorphism (BONNELL & SELANDER, 1984). Whereas for the elephant seals it is clear that the absence of genetic variation is due to a severe bottleneck caused by hunting, it is possible that for the fire salamander (*Salamandra salamandra*) the loss of genetic variation occurs at the border of the distribution area (Veith & Seitz 1990). On the other hand, the cheetah (*Acinonyx jubatus jubatus*) is believed to suffer from the poor genetic variation (O´BRIAN *et al.* 1986).

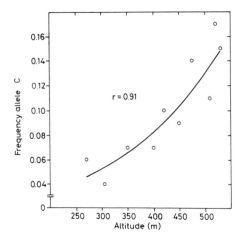

Fig. 5: Correlation between the frequency of an allele of the enzyme Aconitase of the microlepidopteran *Argyresthia mendica* and the elevation above sea level where the populations have been sampled. The sampling stations are in Upper Franconia (FRG). The largest distance between the populations is about 50 km. The topographical arrangement of the populations does not coincide with the increase in gene frequencies. (From SEITZ 1988)

The problem of genetic variation of species is associated with the geographic distribution of genotypes which constitute subspecies, races or ecotypes. Biochemical investigations during the last two decades have revealed numerous cases of otherwise unknown genetic variability which may be the basis for the formation of new species or are the expression of an already completed speciation process. The genetic variability of populations is thought to be closely related to the persistence of populations (SOULÉ M.E. 1987); the loss of genes now is the loss of new species in the future.

This last statement can be illustrated by the large group of phytophagous insects. Many of them are specialists and endangered because of a loss of resources. Some of them are of commercial interest for the biological control of weeds, others because they are pests of many cultivated plants and have therefore been well studied for a long time. It has been shown (BUSH & DIEHL 1982, KOMMA 1990, KLEIN 1991) that these insects show remarkable correlations between the genotypic constitution of their populations and their relation to food resources or geographical distribution which results in a rapid evolution and speciation process. The protection of such species demands the protection of populations in a large variety of ecological and geographical conditions to cover the species with many of its properties.

From these observations, it is clear that species conservation must deal with the problem of genetic variation and try to maintain as many genetic variants as possible.

Perspectives of new genetic techniques

In spite of the successful application of gel electrophoresis of enzymes, this technique is of limited use, in the study of short-term effects. Approximately only one out of 30 mutations can be observed by the electrophoresis of proteins, and enzymes are but a small part of them; it is like looking through a key-hole on the genetic happening. The development of DNA analyses has opened the door for many more detailed studies, even in cases where enzyme electrophoresis showed no variation at all. The discovery of the restriction fragment length polymorphism and its application to population studies (HILL 1987, WETTON et al. 1987, ARCTANDER 1988, BURKE 1989, MENG et al. 1989) was a revolution in accuracy. We can study more loci simultaneously, and consequently follow the way of single genes through a population and between populations.

Besides the use in population biological field studies, DNA analysis can be used for paternity testing in breeding stations of endangered species to select pairs and optimize breeding success. Another application concerns the legal aspect of endangered species. Some species are of high commercial value and are therefore bred (e.g. birds of prey) with more or less success. The high value of these animals gives rise to the illegal removal of eggs or nestlings from eyries, and feigned breeding success. I not caught redhanded, a suspect's quilt is almost impossible to prove. DNA fingerprinting (fig. 6) is an extremely apt tool in such cases. With the exception of identical twins, all individuals differ genetically from each other. Therefore a DNA fingerprint is an unchangeable marker for the identification of individuals. Furthermore, fingerprints are heritable in a Mendelian fashion (WYMAN & WHITE 1980) so that an offspring's fingerprint can only show bands which are found at least in one of the parents.

Our studies (WOLFES *et al.* 1991) have shown that DNA fingerprinting can be successfully applied to all bird species tested up to now. Current experiments in our laboratory show that, besides forensics, DNA fingerprinting is very effective in identifying species that are otherwise difficult to determine.

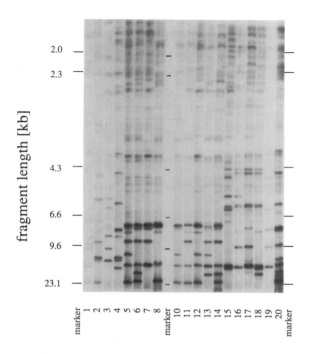

Fig. 6:   DNA fingerprint of eagle owls (*Bubo bubo*). Each numbered lane represents one individual. The DNA was extracted from blood, digested with the restriction enzyme *Hinf*I, separated by gel electrophoresis in agarose, denatured, blotted to a supporting membrane and hybridized with the synthetic oligonucleotide (GGAT)$_4$. The individuals 19 and 20 are the parents of 15 - 18, 5 and 14 are the parents of 6 - 8 and 10 - 13. The individuals 1 - 4 are siblings but unrelated to the other individuals (Photograph provided by J. MÁTHÉ, Mainz).

Conclusions

Population biology can help to develop and perform conservation projects. Scientists have the tools, hopefully they have the power and the courage to protect the subjects of their investigations.

Zusammenfassung

Artenschutz gehört zu den drängendsten und herausforderntsten Problemen biologischer Forschung. Die Poplationsbiologie deckt mit ihren Methoden und dem theoretischen Hintergrund einen großen Teil der Aufgaben ab, die notwendig sind, um Vorschläge für den praktischen Artenschutz zu machen.

Es werden Beispiele vorgestellt, die die Möglichkeiten populationsbiologischer Untersuchungen auf unterschiedlichen Integrationsebenen und mit unterschiedlichen experimentellen oder analytischen Forschungsansätzen zeigen. Die Plastizität der Populationseigenschaften und die wechselseitigen Abhängigkeiten zwischen ökologischen und genetischen Ursachen und Konsequenzen machen es in vielen Fällen schwierig, die Schlüsselfaktoren zu erkennen, die für die Gefährdung von Populationen und Arten verantwortlich sind.

Vom Standpunkt der Ökologie sind die Stochastik der Umwelt und die große Anzahl von biotischen und abiotischen Wechselwirkungen die Kernprobleme, die eine langfristige Vorhersage unmöglich machen. Wenn zusätzlich genetische Eigenschaften betrachtet werden, werden die Schwierigkeiten noch größer. Natürliche und anthropogene Veränderungen der Umwelt verschieben die genetische Struktur von Populationen und verursachen einen Verlust an genetischer Vielfalt. Dies führt zu einer verringerten Anpassungsfähigkeit an eine sich verändernde Umwelt.

Neue analytische Werkzeuge, die ursprünglich für forensische Untersuchungen und Populationsstudien entwickelt worden sind, z.B. DNA-fingerprinting, eröffnen neue Wege, um den Genfluß mit sehr hoher Genauigkeit zu untersuchen. Diese Methoden können auch im praktischen Artenschutz eingesetzt werden, z.B. bei der Feststellung der Elternschaft bei der Zucht und bei der Auswilderung von Tieren, bei der Überprüfung von Populationen, die dem Arterhalt dienen und in der Umweltkriminalistik.

Acknowledgments

This paper was made possible by the financial support of the Bundesministerium für Umwelt, Naturschutz und Reaktorsicherheit, which not only generously sponsored the symposium and printing of this book, but also helped finance the DNA fingerprinting. The Umweltministerium Rheinland-Pfalz (Landesamt für Umweltschutz und Gewerbeaufsicht, Oppenheim) supported the studies on amphibian populations. My students and co-workers kept me busy with ideas and discussions. I thank them all.

References

ARCTANDER P. (1988): Comparative studies of avian DNA by restriction fragment length polymorphism analysis: Convenient procedures based on blood samples from live birds. J. Orn. 129: 205-216.

BONELL M.L. & SELANDER R.K. (1974): Elephant seals: genetic variation an near extinction. Science 184: 908-909.

BURKE T. (1989): DNA fingerprints and other methods for the study of mating success. TREE 4: 139-144.

BUSH G.L. & DIEHL S.R. (1982): Host shifts, genetic models of sympatric speciation and the origin of insect species. Proc. 5th Int. Symp. Insect - Plant Relationships. Wageningen 297-305.

DIAMOND J. (1990): How many species will exist 50 years from now? Verh. Dtsch. Zool. Ges. 83: 221-225.

DOEPNER U. (1990): Migration beim Grasfrosch (*Rana temporaria* L.). Untersuchungen an zwei Populationen in der Saarpfälzischen Moorniederung. Diploma thesis Univ. Mainz.

ECKARTZ-NOLDEN G. (1988): Untersuchungen über die jahreszeitlichen Veränderungen der Phytoplanktonpopulationen des Laacher Sees unter Berücksichtigung der Beziehungen zum Zoooplankton. Diss. Univ. Bonn.

HALBACH U. (1979): Introductory remarks: Strategies in population research exemplified by Rotifer population dynamics. Fortschr. Zool. 25: 1-27.

HILL W.G. (1987): DNA fingerprints applied to animal and bird populations. Nature 327: 98-99.

KLEIN M. (1991): Populationsbiologische Untersuchungen an *Rhinocyllus conicus* FRÖLICH (*Col.; Curculionidae*). Wiss. Verlag Maraun, Frankfurt (Diss. Univ. Mainz).

KOMMA M. (1990): Der Pflanzenparasit Tephritis conura und die Wirtsgattung Cirsium. Wiss. Verlag Maraun, Frankfurt (Diss. Univ. Bayreuth).

MCCAULEY D.E. (1991): Genetic consequences of local population extinction and recolonization. TREE 6: 5-8.

NEVO E., BEILES A. & BEN-SHLOMO R. (1984): The evolutionary significance of genetic diversity: ecological, demographic and life history correlates. Lect. Notes in Biomath. 53: 13-213.

O'BRIAN S.J., WILDT D.E., GOLDMAN D., MERRIL C.R. & BUSH M (1983): The cheetah is pauperate in genetic variation. Science 221: 459-462.

POETHKE H.J., OERTEL D. & SEITZ A. (1991): Risk assessment of toxicants to pelagic food-webs: A simulation study. In: MOELLER, D.P.F. (Ed.): Advances in Systems Analysis. Springer Verlag (in press).

REH W. & SEITZ A. (1990): Influence of land use on the genetic structure of populations of the Common Frog (*Rana temporaria*). Biol. Cons. 54: 239-249.

SEITZ A. (1988): Microgeographic Variation of Genetic Polymorphism in *Argyresthia mendica* (Lep.: Argyresthiidae). In: G. DE JONG (ed.) Population Genetics and Evolution. Springer Verlag Heidelberg 202-208.

SEITZ A. & HALBACH U. (1973): How is the population density regulated? Experimental studies on rotifers and computer simulations. Naturwiss. 60: 51.

SEITZ A. & KOMMA M. (1984): Genetic polymorphism and its ecological background in Tephritid populations (Dipt.: Tephritidae). In: WÖHRMANN K. & LOESCHCKE V. (ed.) Population biology and evolution. Springer Verlag Heidelberg 143-158.

SHAW C.R. & PRASARD R. (1970): Starch gele electrophoresis of enzymes —a compilation of recipies. Biochem. Genet. 4: 297-320.

SMITHIES O. (1955): Zone electrophoresis in starch gels: group variations in the serum proteins of normal human adults. Biochem. J. 61: 629-641.

SOULÉ M.E. (1987): Viable populations for conservation, Cambridge University Press.

VEITH M. & SEITZ A. (1990): Population genetical and morphological investigation on the Fire Salamander (*Salamandra salamandra* L.) in Israel: effects of an extreme environment. Verh. Dtsch. Zool. Ges. 83: 493-494

WETTON J.H., CARTER R.E., PARKIN, D.T. & WALTERS D. (1987): Demographic study of a wild House Sparrow population by DNA fingerprinting. Nature 327: 147-149.

WOLFES R., MÁTHÉ J. & SEITZ A. (1991): Forensics of birds of prey by DNA fingerprinting with ^{32}P-labeled oligonucleotide probes. Electrophoresis (in press).

WYMAN A.R. & WHITE R. (1980): A highly polymorphic locus in human DNA. Proc. Nat. Acad. Sci. USA 77: 6754 - 6758.

ZWÖLFER H. (1979): Strategies and counter strategies in insect population systems competing for space and food in flower heads and plant galls. Fortschr. Zoolog. 25, 2/3: 331-353.

Genetics and Conservation Biology

A. R. Templeton, Department of Biology, Washington University, St. Louis, MO 63130, USA

Abstract

Genetics can be applied to many problems in the area of conservation biology, and four such uses will be illustrated in this paper. First is conservation forensics in which genetic techniques are used to aid the enforcement of laws concerning endangered species. An example of this application is the use of DNA "fingerprinting" to identify the geographical source of tusks as part of an effort to restrict poaching of African elephants. A second application is in systematics. Before a conservation program can be designed, we obviously need to know what it is we are trying to conserve. Genetic techniques are emerging as a major tool in systematics and have proven to be useful in identifying taxa that behave as independently evolving genetic lineages (i.e., species). An example of this is provided by studies on wild cattle species. The cattle also illustrate a third application of genetics in conservation: the detection and monitoring of hybridization. Hybridization can cause genotypic extinction through extensive gene flow, and the incidence of hybridization has been greatly increased by human activities through the introduction of exotics and the disturbance of natural environments and isolating barriers. The final area is genetic management of natural and captive populations of endangered species. Genetic management is necessary both for the short-term health of the species (e.g., inbreeding depressions) and the long-term adaptive flexibility of the species (e.g., preserving genetic diversity). Examples of genetic management of both captive and natural populations will be given.

Ecology is the biological science that is most commonly linked to conservation issues in the public mind. However, the science of genetics is also a major contributor to conservation biology. The purpose of this paper is to outline four major applications of genetics to conservation issues: 1) conservation forensics, 2) systematics, 3) hybridization, and 4) the management of natural and captive populations. Examples of these applications will be given from work being carried out in my laboratory.

Conservation Forensics

The science of genetics is having a major impact on criminal forensics, primarily by using genetic screening to provide individual identification. Just as there are criminal and civil laws, there are also conservation laws. The effective enforcement of many of these laws requires the identification of individuals or tissues. Modern genetic techniques, such as DNA fingerprinting, are ideal for this purpose.

To perform a DNA fingerprint, total cellular DNA is isolated and cut with a restriction enzyme, usually a 4-base cutter. The cut DNA is run through an electrophoretic gel to separate the DNA fragments by size. Specific fragments are identified by hybridizing the DNA in the gel to radioactive probes which will only hybridize to the cellular DNA fragments to which they are homologous. With DNA fingerprinting, a specific class of probes are used called "mini-satellite" probes. These probes consist of a highly conserved core sequence which will hybridize with homologous sequences in virtually all eukaryotic organisms (TEMPLETON *et al.*, 1990; ROGSTAD *et al.*, 1988, 1989). Because of this highly conserved core sequence, DNA fingerprinting is extremely portable from one species to another, making it particularly useful for conservation forensics, an area dealing with a multitude of diverse species of plants and animals.

Although the core sequence is highly conserved, minisatellite probes are able to detect extremely high levels of genetic variation because they are commonly found in variable places in the genome, and at each location they tend to be found in tandem clusters with variable numbers of repeats. Accordingly, this system taps into much intraspecific variation, which makes it an ideal system for providing individual identification.

Drs. John Patton and Nicholas Georgiadis are applying this technique to a legal problem in conservation: enforcement of poaching bans on African elephants. Poaching for ivory is causing the African elephant (*Loxodonta africana*) to decline so rapidly that few populations are expected to persist beyond the next decade unless the poaching can be severly curtailed. In an extreme measure designed to halt poaching, the member countries of the Convention on International Trade in Endangered Species voted to ban all trade in ivory after much debate. The debate concerns a fundamental dispute between the countries north of Zimbabwe versus the southern countries of Zimbabwe, South Africa and Botswana. Despite the fact that all hunting of elephants is illegal in the northern countries, the elephant herds there are rapidly going to extinction due to poaching. In contrast, carefully regulated culling in Zimbabwe has shown that elephant products could constitute one of the most profitable "extractable resources" from large areas that are not protected as parks or reserves. As a consequence, trade in ivory from legally culled elephants provides a strong economic incentive for the conservation of both elephants and their habitats. Indeed, the elephant herds in Zimbabwe have actually increased in size under the controlled culling policy. Accordingly, the southern countries have argued that controlled culling is the best long-term solution to elephant and elephant habitat conservation. These countries regard the total ban as only a short-term solution to aid the northern countries in getting control over their poaching problems, and have agreed to the ban only under the condition that the question be reconsidered in 1991. In the meantime, they are continuing to cull their herds in anticipation of the trade resuming some day. This brings up the central dilemma: how to distinguish the ivory culled from the well-managed southern herds from the illicit ivory obtained by poaching in the northern herds?

Patton and Georgiadis are attempting to see if DNA fingerprinting can provide an answer to this question. They obtained samples from 100 tusks confiscated from poachers in Kenya and Tanzania. Many of these tusks had attached bits of dried tissue, and this tissue contains high quality DNA that yields interpretable DNA "fingerprints". Moreover, they found a marker that differentiates between Kenya and Tanzania, so it is hoped that markers will also be found to differentiate the more geographically separated populations from the north and the south.

An alternative approach is to use the polymerase chain reaction (PCR) to amplify specific segments of DNA in the dried tissue. With PCR, one uses small primer sequences of highly conserved DNA to amplify a small piece of DNA found in the cellular DNA. This

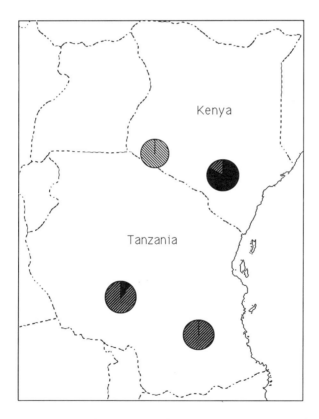

Fig. 1:   The distribution of the *Sau*3A restriction site polymorphisms of elephant mitochondrial DNA in eastern Africa. There are three haplotypes defined by this enzyme in a PCR ampflified 2 kb segment of the mitochondrial DNA. Pie diagrams indicate the haplotype frequencies at various collecting sites in Kenya and Tanzania.

technique is particularly useful for dried tissue samples, in which the DNA is already highly degraded. PATTON and GEORGIADIS have successfully amplifed segments of both nuclear and mitochondrial DNA from tusks. Once amplified, the DNA can be screened for variation either through sequencing or cutting with restriction enzymes. So far, they have found *Sau*3A restriction site variation in one *mt*DNA segment that indicates much geographical subdivision within elephants (Figure 1). Three elephants from western Kenya had a unique variant found nowhere else. The eastern Kenyan populations have one common variant (10/12) that is rare in Tanzania (1/13) and another variant that is rare in eastern Kenya (2/12) but common in Tanzania (12/13). These initial results strongly indicate that it should be possible to identify the country of origin of a tusk based on genetic screening. If so, this will be a valuable tool in elephant conservation and could contribute to the resolution of the debate between the northern versus southern African countries.

Systematics

Before we can design a conservation program, we need to know what it is we are trying to conserve. Accordingly, it is essential that we have basic systematic information on endangered taxa. Genetics has emerged as a major tool in systematics that is particularly useful because the same basic techniques and methods of data analyses are applicable to virtually all species. This portability is critical because we often do not have the time to perform a detailed systematic study of many endangered groups of plants and animals based on their morphology, life history, development, and ecology. Genetic techniques can be applied much more readily to most groups of organisms, and the resulting systematic inferences are often more accurate than those based on morphology or other types of non-genetic data (WILSON *et al.*, 1985).

Their are many techniques and methods of analysis for making such systematic inferences from genetic data (TEMPLETON, 1986a), but they will not be given here. Basically, by studying homologous pieces of DNA (through restriction mapping, sequencing, etc.) and observing how these homologous pieces vary among taxa, it is possible to define the taxa that are behaving as independently evolving lineages (i.e., species for purposes of this paper) and their evolutionary relationships to one another.

We have performed a genetic systematic study of the mammalian subfamily Bovinae, which includes the domestic cattle and their wild relatives. These taxa are morphologically, behaviorally, and ecologically distinct from each other, but many of the taxa in this family can and do hybridize (e.g., domestic cattle can hybridize to at least some extent with all other *Bos* and *Bison* species). Because many of these taxa do not satisfy the criterion of species status under the biological species concept (MAYR, 1963), questions of species and even genus status abound in this subfamily. To address this issue, (WALL *et al.*, 1991) constructed a molecular phylogeny of the Bovinae using restriction maps of nuclear ribosomal DNA (rDNA). The resulting maximum parsimony phylogeny is shown in Figure

2. As can be seen, the genus *Bos* is paraphyletic with respect to the genus *Bison*, and there are no detected genetic differences in the rDNA among the domestic cattle species and the two *Bison* species. Hence, these two genera should probably be combined and further genetic studies will be needed to resolve the relationships among the domestic cattle species and the *Bison* species. However, the other wild cattle species in the genus *Bos* are behaving as separate evolutionary lineages from domestic cattle. Thus, we conclude that these wild cattle species are worthy of conservation efforts as distinct species, and not merely as wild varities of domestic cattle.

As this example shows, genetic surveys are a valuable tool in systematics. Such surveys provide a portable methodology for identifying taxa that have species status and for identifying taxa that warrant further investigation. Such basic taxonomic information is critical in formulating conservation programs.

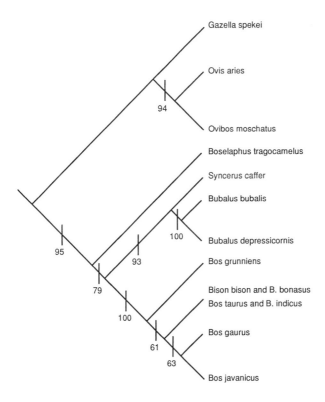

Fig. 2:   The maximum parsimony cladogram of the subfamily Bovinae based on ribosomal DNA restriction site data. The numbers refer to specific restriction sites that define the clades.

Hybridization

The fact that domestic cattle can hybridize with all other wild cattle species to at least some extent illustrates another major problem in conservation: hybridization. Hybridization has long been recognized as a common phenomenon in plants. As a consequence, Botanists have recognized many species that exist in larger units known as syngameons, which (GRANT 1981) defines as "the most inclusive unit of interbreeding in a hybridizating species group." More recently, it has been realized that syngameons are also extremely common in animals (TEMPLETON, 1989), as illustrated above by *Bos/Bison*.

Although the species within a syngameon are poorly defined by reproductive isolating barriers, they generally maintain sharply defined boundaries by morphological, ecological and genetic criteria (TEMPLETON, 1989); once again, as already illustrated by *Bos/Bison*. However, botanists have long appreciated a major exception to the above "in general" statement: disturbed environments. Species within a syngameon are often maintained as distinct lineages by their ecological distinctiveness and/or biogeographical constraints. Disturbance of the environment can break down the ecological and biogeographical factors that maintain the species and can result in the genetic extinction of the species through hybridization. For example, two sculpin species, *Cottus confuses* and *C. cognatus*, are morpholocially, ecologically, and genetically distinct. Their breeding seasons normally do not overlap, resulting in complete reproductive isolation. However, the construction of a dam altered the thermal environment, causing an overlap of breeding season which lead to the replacement of both species by a hybrid swarm in the area by the dam (ZIMMERMAN & WOOTEN, 1981). Another major form of disturbance is the introduction of domestic or non-native species to new areas. For example, the rock dove (*Columba livia*) is undergoing genetic extermination in those parts of Europe where it is in sympatry with feral pigeons, its domestic but genetically very distinct descendant (JOHNSTON *et al.*, 1988).

Another species that is endangered by hybridization is the banteng (*Bos javanicus*), a wild cattle species. In 1900, banteng were widely distributed in the primary forests which still covered most of the island of Java at that time. By 1960, the natural forest had largely disappeared and what little remained was highly fragmented. By this time, only about 1000 banteng remained, also distributed in a highly fragmented fashion. Conservation efforts were initiated and have stabilized the population at at least 700 animals and perhaps more than a 1000 (ASHBY & SANTIAPILLAI, 1986). However, many of these populations have been or currently are exposed to domestic cattle. For example, it is known that three females of *Bos indicus* were introduced in 1922 to the Penanjung Pangandaran Nature Reserve, which currently contains a population of about 60 banteng. Hybridization occurred, and the managers of the reserve have selected on this herd so that the present population looks like banteng. However, the extent of genetic contamination in this herd is unknown, and because of the selection, the morphological data are useless in addressing this issue. Similarly, most of the other banteng populations in Java are or recently have been at risk for hybridization with domestic cattle (ASHBY & Santiapillai, 1986). In addition to natural populations of banteng, captive herds have also been established as a back-up to the nature reserves. However, the genetic purity of many of these captive populations has also been questioned (DAVIS & READ, 1985). Consequently, the primary threat to the continued existence of the genetic lineage known as banteng is not a numerical decline, but hybridization with the genetically distinct lineages of domestic cattle.

Hybridization is a genetic problem, and hence genetics is the primary tool needed to assess this problem. For example, the status of the captive herds of banteng in North America have been assessed by performing genetic surveys on these herds, domestic breeds, and known hybrids that revealed many diagnostic markers that clearly distinquish *Bos javanicus* from the two domestic species, *Bos taurus* and *Bos indicus*. These genetic surveys indicate that the captive herd of banteng in North America is not genetically contaminated, and hence represents a legitimate back-up to conservation efforts to preserve this species (DAVIS & READ, 1985; DAVIS *et al.*, 1988). The next step in banteng conservation is to perform genetic surveys on the natural populations to assess directly the impact of hybridization in nature.

Genetic Management

Once a species has been identified and the problem of hybridization has been shown to be irrelevant or otherwise dealt with, genetics remains an essential part of a species conservation program for three principle reasons. First is genetic husbandry. The preservation of captive or natural populations of endangered species usually requires long-term, multigenerational management programs. Such long-term programs must not only deal with the immediate horticultural or husbandry needs of captive populations and the habitat requirements of natural populations, but also with inbreeding or outbreeding depressions that frequently occur in multi-generational managment programs and that can seriously endanger the long-term viability of a population. Both of these types of fitness depression are related to the genetic environment (the amount of individual heterozygosity and homozygosity, the stability of complexes of genes at several loci, etc.), which in turn is determined by system of mating, population size, and population subdivision. When a population is brought into captivity or the population structure of a natural population is altered by human activities (e.g., habitat fragmentation), one or more of these factors can be changed, resulting in a new genetic environment. This altered genetic environment induces a state of maladaptation. Populations of endangered species must be managed to either avoid altering the genetic environment or to adapt them to the alterations. Hence, genetic management is an essential aspect of husbandry in multi-generational management programs.

A second reason for genetic management is to preserve adaptive flexibility. The environments and habitats of many species are being altered in many ways be human activities, and we can expect these alterations to continue. Because we cannot foresee the future, the long-term viability of a species will often depend upon its adaptive flexibility, which in turn depends critically upon its level of genetic variability. Hence, populations must also be managed to preserve genetic variation.

Third, the genetic variation present in a species is a valuable biological resource in this era of genetic engineering. Agricultural scientists have long used wild relatives of domestic plants and animals as a source of genes to improve our domestic stocks. With recombinant DNA technology, a gene from any species, even if it is not evolutionarily close to a domestic species, represents a potential resource for improving our agricultural stocks. Hence, the genetic diversity found in the living world represents a valuable natural resource. This genetic diversity represents an historical legacy that can be squandered and lost, and hence it must be managed carefully.

These general considerations lead to two primary genetic goals in species management: 1) preserving genetic diversity, and 2) ideally altering the genetic environment of the species as little as possible or, less ideally, adapting the species to alterations that are forced by practical constraints. Two examples of genetic managment will now be given; one for natural populations, and the other for a captive population.

The first example concerns efforts by my laboratory and the Missouri Conservation Commission to reintroduce populations of the collared lizard, *Crotaphytus collaris*, on restored habitats in the Missouri Ozarks. These lizards live on Ozark glades: rocky, treeless outcrops usually near the tops of ridges with southerly or southwesterly exposures. These glades provide xeric habitats separated from one another by the predominant oak-hickory forest found in the Ozarks. Without periodic forest fires, the forest will often gradually encroach upon the glade. Because of human intervention to prevent forest fires, many glade habitats have been destoryed or seriously reduced in area. The Missouri Conservation Commission has been restoring many of these habitats and then maintaining them by periodic burns. However, the collared lizard, which is near the top of the glade food chain, has undergone local extinction in many of these glades. Accordingly, it was decided to capture lizards from natural populations and reintroduce them into the empty glades. The question is: how to do this introduction in such a manner as to preserve genetic variation?

To create a sampling design to meet this genetic goal, we had to know how genetic variation is distributed within and among the natural populations of this lizard. Hence, we performed genetic surveys on these lizards that revealed that most of the genetic diversity is present as differences between subpopulations inhabiting different glades, and very little is present in the form of individual heterozygosity, as shown in Figure 3 for allozymes and restriction site variation in mitochondrial and nuclear ribosomal DNA (from TEMPLETON *et al.*, 1990). Unfortunately, these genetic survey techniques revealed low overall levels of variation. In order to increase our genetic resolution and thereby strengthen our inferences about genetic subdivision, we performed an additional genetic survey on five glade populations scattered throughout the eastern Missouri Ozarks (details on sample sizes and geographical location are given in (TEMPLETON *et al.*, 1990), with the addition of five lizards to the Proffit Mountain sample) by using total cellular DNA cut with the restriction enzyme *Hae*III and probed with the human fingerprinting clone 33.15 (kindly provided by Dr. A. JEFFRIES). We found a total of 13 variable, high molecular weight bands that could be reliably scored (there was obvious variation at some low-molecular weight bands as well, but it was difficult to score). In order to quantify the extent of population subdivision from DNA fingerprinting data, the sample reuse algorithm described in (DAVIS *et al.*, 1990) was modified to calculate the proportion of shared bands (S) that randomly drawn individuals from the same and different glades have in common with an index individual. Because allelic homologies are not known, standard F statistics cannot be calculated, but it is possible to calculate the coefficient of relationship among individuals within glades relative to that between glades as $r = [S_{(within)} - S_{(between)}]/[1 - S_{(between)}]$.

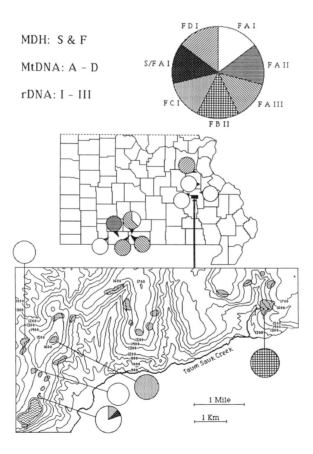

Fig. 3:   Distribution of isozyme, mitochondrial DNA, and nuclear ribosomal DNA genetic variants four collecting sites (data are given in Table 3 of TEMPLETON *et al.*, 1990). An expanded scale maj reas on this map. Contour intervals are 100 feet.

Since band phenotypes are additive, $r = 2F_{st}/(1+F_{is})$ (ROTHMAN et al., 1974). Although F_{st} cannot be estimated from r, note that $F_{st} = 0$ if and only if r = 0. Hence, the null hypothesis of no population subdivision ($F_{st} = 0$) can be tested with r. Moreover, as subdivision increases, so does r. Hence, r is a legitimate and meaningful indicator of population subdivision.

The estimated r for these five glade populations is 0.44 with a 95% confidence interval of 0.32 to 0.54. This is a very large value of r and indicates a high degree of genetic subdivision among these populations. Indeed, most glades have bands that are unique to them, including glades only 2 km apart. These results and Figure 3 illustrate that a sample of lizards taken from a single glade subpopulation would miss most of the genetic diversity present in the species regardless of the size of the sample. The best way to preserve this species' genetic diversity is to sample a few animals from many different glades rather than a large number of animals from a few glades.

This example illustrates the critical importance of performing genetic surveys on endangered species. Simply establishing a population from a large number of founders does not insure that genetic diversity is being preserved. Without a knowledge of how the species' genetic diversity is divided within and between local populations, it is impossible to design a sampling program that will preserve a substantial portion of the species' genetic diversity.

Genetic surveys are also useful in dealing with the genetic environment problem. The best way to avoid inbreeding or outbreeding depressions is to minimize changes in the genetic environment under the management program. To do this we need to know what that environment is. This knowledge can be acquired from genetic surveys on the natural populations and/or by making observations on the breeding system and life history of the species (TEMPLETON, 1990). Thus, genetic surveys on natural populations play a doubly important role in designing management programs.

In the case of the collared lizards, the genetic survey reveals a highly subdivided population structure characterized by high levels of individual homozygosity within local breeding populations. This type of genetic environment could be preserved by capturing all the lizards for release on a single restored glade from a single natural glade. Genetic diversity could be preserved at the global level by using a different natural glade population as the source for each new release. However, as is so common in conservation programs, practical constraints emerge that force us to make some compromises. In the case of the collared lizards, very few glades have large enough natural populations to sustain the harvesting level necessary for a reintroduction (we use 8 to 10 animals per release). In order not to endanger these natural populations, we were forced to adopt the rule that no more than two animals would be taken from any single natural population. This rule necessitates mixed releases on the restored glades. Consequently, the populations on the restored glades will experience a greatly altered genetic environment (at least for the first few generations) characterized by high levels of individual heterozygosity and recombination between previously homozygous blocks of genes. These conditions set up the possibility of an outbreeding depression that may endanger the viability of the released populations (TEMPLETON, 1986b). However, experimental results (ANNEST & TEMPLETON, 1978) indicate that the brunt of the outbreeding depression occurs in the F_2 and backcross generations, and if that "fitness bottleneck" can be passed, the resulting population will be as or more viable than the original parental populations (TEMPLETON *et al.*, 1990). Hence, it is critical to monitor the released populations for the first few generations in order to see if an outbreeding depression is occurring and if so, to decide if some additional intervention (e.g., the release of more animals) is necessary to get the population through this fitness bottleneck.

In order to do this monitoring, we took blood samples for genetic analysis from all lizards before releasing them. The genetic markers provided by these surveys will allow us to identify F_2 and backcross individuals in the released populations, and hence will allow a direct study of outbreeding depression in these populations. Because of the lizards' long generation time, the oldest released populations are just beginning to enter into the F_2 and backcross generations, so we currently do not know if outbreeding depression will be a problem. However without a monitoring program in place, we would never know the answer to this important management question. Hence, this monitoring will be particularly important in the next few years. Moreover, these genetic markers will allow us to obtain information on many demographic features of the released populations. For example, we know that male lizards are territorial during mating season. This may imply that only a few males actually reproduce, and this in turn may imply that the optimal release composition should be skewed in favor of females. By genetically monitoring our released populations, we can directly determine the number of reproducing males.

We hope to demonstrate by monitoring the collared lizard populations that genetic surveys are a useful follow-up tool for both genetic and demographic management. Conservation biology is often a crisis science in which we need to act before all the relevant facts are known. Consequently, using genetic surveys as a monitoring tool allows us to learn both from our successes as well as our failures.

Another example of the importance of practical constraints in genetic management is provided by a captive herd of Speke's gazelle (*Gazella spekei*). The natural habitat of this species is the border region between Ethiopia and Somalia. Because of prolonged warfare in that region, it has not been feasible to monitor genetic variation in the natural population or to infer the natural genetic environment. However, studies on the response of captive breed animals to inbreeding and outbreeding imply that the natural populations of this species are highly outbred and heterozygous (TEMPLETON, 1987). Moreover, genetic surveys on the captive population revealed that the original founding animals had much genetic polymorphism and high levels of individual heterozygosity (TEMPLETON *et al.*, 1987). Hence, ideally the captive herd should be managed so as to avoid inbreeding by preferentially breeding unrelated individuals.

Unfortunately, the avoidance of inbreeding is impractical in this case. The founding population consisted of one male and three females, and no other animals were available at the time. Once the original founders were no longer reproducing, all possible matings in this herd had to be between close relatives. Thus, there was no option other than to adapt the captive population to high levels of inbreeding. This adaptation to inbreeding was particularly critical for the Speke's gazelle because the initial inbreeding depression was severe, with about 80% of liveborn individuals dying before maturity. The details of this adaptation to inbreeding are given elsewhere (TEMPLETON & READ, 1983, 1984); here, we only note that this program successfully reduced the inbreeding depression so it no longer threatened the maintenance of the herd. This example clearly shows that inbreeding depression can be managed.

With the inbreeding depression reduced, we now addressed the other genetic goal: preserving genetic diversity. Instead of maintaining the gazelles as one large breeding group, we implemented a subdivided population structure by establishing subherds at different zoos throughout the United States and limiting transfer of animals between the subherds as much as possible. This subdivided herd strategy is extremely effective in maintaining genetic diversity for long periods of time with only a small number of individuals (TEMPLETON, 1990), but it can only be implemented when inbreeding depression is not a serious husbandry issue since subdivision results in high levels of inbreeding.

With both the lizards and the gazelles, compromises had to be made between our ideal program and what was practical. In both cases, our compromise favored the preservation of genetic diversity over the stability of the genetic environment. The preservation of genetic diversity should have priority because it allows the species to retain its adaptive flexibility and allows the genetic manager to let evolution help in preserving the species, as with the elimination of inbreeding or outbreeding depressions. If genetic variability is lost, the chances of long-term survival of the species are reduced and options for genetic management are eliminated. We therefore conclude that some rules of conservation biology, such as "the avoidance of inbreeding", are of less importance than the preservation of genetic diversity, and our management programs should be shaped accordingly.

Unfortunately, the science of conservation biology is already being burdened by a number of rules which are often blindly applied. For example, one rule for the preservation of genetic diversity is the 50/500 rule, which states that founding populations should consist of 50 breeding individuals, and that the long-term breeding size should be brought up to 500 individuals. Both the collared lizard and Speke's gazelle programs seriously violate this rule. Yet in both cases, genetic surveys indicate that high levels of genetic diversity are indeed being preserved. To understand this apparent contradiction, one needs to look at the underlying evolutionary arguements used to justify the 50/500 rule. This rule was based on the assumption that the natural population is large and panmictic, an assumption that is seriously violated in the case of the collared lizards. It is also based on the assumption that the managed population is panmictic and not subdivided, an assumption that is seriously violated in the case of the Speke's gazelles. Consequently, the 50/500 rule is completely irrelevant to both of these cases. What is important is not the rule, but the underlying evolutionary arguements that were used to generate the rule in the first place. If the evolutionary underpinning of the rule is well understood, then that basic evolutionary knowledge can be applied to specific cases to yield recommendations that will frequently violate the general "rule". A management program should not be designed around a set of pre-established "rules"; rather it must be designed through the creative application of basic evolutionary and genetic principles to specific instances.

Zusammenfassung

In der öffentlichen Meinung wird aus den biologischen Disziplinen am ehesten die Ökologie mit Naturschutzfragen in Verbindung gebracht. Jedoch leistet auch die Genetik einen wichtigen Beitrag zum Naturschutz. Der Zweck der vorliegenden Arbeit liegt darin, vier Hauptanwendungen der Genetik zu Naturschutzproblemen zu umreißen: 1) Rechtsmedizinische Methoden im Artenschutz, 2) Systematik, 3) Hybridisierung und 4) Management von natürlichen und in Gefangenschaft gehaltenen Populationen. Als Beispiele für diese Anwendungen werden Arbeiten angeführt, die in meinem Labor durchgeführt wurden.

Im ersten Beispiel wird die Anwendung rechtsmedizinischer Methoden im Artenschutz gezeigt, mit denen die Durchsetzung von Gesetzen zum Schutz bedrohter Arten unterstützt wird: Die Anwendung des "DNA-Fingerprintings" (genetischer Fingerabdruck) zur Bestimmung der geographischen Herkunft von Stoßzähnen als Beitrag bei den Bemühungen, die Wilderei des Afrikanischen Elefanten einzudämmen. Eine zweite Anwendung liegt in der Systematik. Bevor ein Artenschutzprogramm entwickelt werden kann, müssen wir zunächst wissen, was genau geschützt werden soll. Genetische Methoden werden zu einem Hauptwerkzeug der Systematik und erwiesen sich nützlich bei der Identifizierung von Taxa, die sich als unabhängig entwickelnde genetische Linien (z.B. Arten) verhalten. Studien an Wildrindern sind hierfür ein Beispiel. Dabei kann eine dritte Anwendung der Genetik im Artenschutz illustriert werden: der Nachweis und die Verfolgung von Hybridisierungen. Hybridisierung kann durch extensiven Genfluß zum genotypischen Aussterben führen. Das Auftreten von Hybridisierungen hat durch menschliche Aktivitäten infolge der Einführung exotischer Arten und Störungen der natürlichen Umwelt und der Isolationsschranken stark zugenommen. Schließlich ist das genetische Managemaent von natürlichen und in Gefangenschaft gehaltenen Populationen bedrohter Arten eine weiteres Gebiet der Genetik im Artenschutz. Genetisches Management ist notwendig zum einen zur kurzfristigen Gesunderhaltung von Arten (z.B. Vermeidung von Inzuchtdepression) und zum anderen zur langfristigen Erhaltung der Anpassungsfähigkeit von Arten (z.B. Erhaltung genetischer Vielfalt). Beispiele für das genetische Management von sowohl natürlichen als auch in Gefangenschaft gehaltenen Populationen werden vorgestellt.

Acknowledgements

This work was supported by NIH grant R01 GM31571.

References

ANNEST, L. & A.R. TEMPLETON. (1978): Genetic recombination and clonal selection in *Drosophila mercatorum.* Genetics 89:193-210.

ASHBY, K. R. & C. SANTIAPILLAI. (1986): The status of the banteng (*Bos javanicus*) in Java and Bali. WWF/IUCN Report 28.

DAVIS, S.K. & B.M. READ. (1985): The status of North American captive herds of the banteng (*Bos javanicus* d'Alton). Zoo Biology 4: 269-279.

DAVIS, S.K., B. READ, & J. BALKE. (1988): Protein electrophoresis as a management tool: detection of hybridization between banteng (*Bos javanicus* d'Alton) and domestic cattle. Zoo Biology 7: 155-164.

DAVIS, S.K., J.E. STRASSMAN, C. HUGHES, L.S. PLETSCHER, & A.R. TEMPLETON. (1990): Population structure and kinship in *Polistes* (Hymenoptera, Vespidae): An analysis using ribosomal DNA and protein electrophoresis. Evol. 44: 1242 - 1253.

GRANT, V. (1981): Plant Speciation, 2nd Edition. Columbia University Press, New York.

JOHNSTON, R.F., D. SIEGEL-CAUSEY, & S.G. JOHNSON. (1988): European populations of the rock dove *Columba livia* and genotypic extinction. Am. Midland Nat. 120:1-10.

MAYR, E. (1963): Animal Species and Evolution. Harvard University Press, Cambridge, MA.

ROGSTAD, S.H., J.C. PATTON II, & B.A. SCHAAL. (1988): M13 repeat probe detects DNA minisatellite-like sequences in gymnosperms and angiosperms. Proc. Natl. Acad. Sci. USA 85: 9176-9178.

ROGSTAD, S.H., J.C. PATTON II, & B.A. SCHAAL. 1989): A human minisatellite probe reveals RFLPs among individuals of two angiosperms. Nuc. Acids Res. 16:11378.

ROTHMAN, E.D., C.F. SING, & A.R. TEMPLETON. (1974): A model for the analysis of population structure. Genetics 78: 943-960.

TEMPLETON, A.R. (1986a): Relation of humans to African apes: a statistical appraisal of diverse types of data. In Evolutionary Processes and Theory, ed. S. Karlin and E. Nevo, pp. 365-388. Academic Press, New York.

TEMPLETON, A.R. (1986b): Coadaptation and outbreeding depression. In Conservation Biology: Science of Scarcity and Diversity, ed. M. Soule, pp. 105-116. Sinauer, Sunderland, Massachusetts.

TEMPLETON, A.R. (1987): Inferences on natural population structure from genetic studies on captive mammalian populations. In Mammalian Dispersal Patterns, ed. B.D. Chepko-Sade and Z.T. Halpin, pp. 257-272. University of Chicago Press, Chicago.

TEMPLETON, A.R. (1989): The meaning of species and speciation: a genetic perspective. In Speciation and its Consequences, ed. D. Otte and J.A. Endler, pp. 3-27. Sinauer, Sunderland, Massachusetts.

TEMPLETON, A.R. (1990): Offsite breeding of animals and implications for plant conservation strategies. In Genetics and Conservation of Rare Plants, ed. K. Falk and K. Holsinger, Oxford University Press, New York, in press.

TEMPLETON, A.R., S.K. DAVIS, & B. READ. (1987): Genetic variability in a captive herd of Speke's gazelle (*Gazella spekei*). Zoo Biology 6: 305-313.

TEMPLETON, A.R. & B. READ. (1983): The elimination of inbreeding depression in a captive herd of Speke's gazelle. In Genetics and Conservation: A Reference for Managing Wild Animal and Plant Populations, ed. C.M. Schonwald-Cox, S.M. Chambers, B. MacBryde, and L. Thomas, pp. 241-261. Addison-Wesley, Reading, Massachusetts.

TEMPLETON, A.R. & B. READ. (1984): Factors eliminating inbreeding depression in a captive herd of Speke's gazelle. Zoo Biology 3:177-199.

TEMPLETON, A.R., K. SHAW, E. ROUTMAN, & S.K. DAVIS. (1990): The genetic consequences of habitat fragmentation. Ann. Missouri Bot. Gard. 77: 13-27.

WALL, D.A., S.K. DAVIS, & B.M. READ. (1991): Evolutionary relationships in the subfamily Bovinae (*Mammalia: Artiodactyla*): An analysis using ribosomal DNA. J. Mamml., in press.

WILSON, A.C., R.L. CANN, S.M. CARR, M. GEORGE, U.B. GYLLENSTEIN, K.M. HELM-BYCHOWSKI, R.G. HIGUCHI, S.R. PALUMBI, E.M. PRAGER, R. D. SAGE, & M. STONEKING. (1985): Mitochondrial DNA and two perspectives on evolutionary genetics. Biol. J. Linn. Soc. 26: 375-400.

ZIMMERMAN, E.G. & M.C. WOOTEN. (1981): Allozymic variation and natural hybridization in sculpins, *Cottus confuses* and *Cottus cognatus*. Biochem. Syst. Ecol. 9: 341-346.

Species Conservation: A Population-Biological Approach
A. SEITZ & V. LOESCHCKE (eds.) © 1991 Birkhäuser Verlag, Basel

Gene Conservation and the Preservation of Adaptability

H-R. Gregorius,   Abteilung für Forstgenetik und Forstpflanzenzüchtung, Georg-August Universität, Büsgenweg 2, D-3400 Göttingen, Germany

Abstract

Preservation of the adaptability of a resource population is emphasized as an indispensable objective of gene conservation. Therefore, methods of dynamic conservation deserve priority. Adaptability requires the existence of genetic variation that is either adaptively inferior or neutral under the respectively prevailing environmental conditions. On this basis, the significance of genetic load and environmental heterogeneity for the maintenance of genetic variation in gene resource populations is discussed, and some elementary principles to be considered in gene conservation are pointed out. It is suggested that criteria for the determination of resource population sizes be oriented at the loss of genetic variation that can be tolerated over a specified number of generations as a consequence of random genetic drift. A model taking account of this aspect is presented, and the population sizes resulting for different systems of reproduction are computed. The relationships between dynamical gene conservation and population (species) protection are emphasized, and the necessity for ecosystem protection is argued.

Introduction

By far the majority of activities in the conservation of genetic resources originated from and are still dominated by agricultural needs, which explains the preponderance of breeding objectives in the conceptual treatment of the subject as well as in the declaration and management of gene resources. One of the consequences of this situation is that problems of static conservation and of the identification and protection of reserves receive most of the attention (see e.g. FORD-LLOYD and JACKSON 1986). In fact, preserving genetic variants of known or prospective value for human use is the driving force of this gene conservation[*] perspective.

---

[*]   If not stated otherwise, the term "gene conservation" is used in this paper to include conservation objectives directed towards both genes and genotypes.

A different conservation perspective is concerned with the conditions for survival of endangered populations or species irrespective of their actual or potential use. Here, aspects of dynamic conservation *in situ* have priority, and the support of mechanisms improving stability and persistence of the associated ecosystems is the objective of management (for a review see e.g. SCHONEWALD-COX et al. 1983). Since dynamic processes usually imply changes in the composition of populations, conservation measures are aimed at maintaining at least approximately steady (equilibrium) states. Because such efforts can only be limited to a small number of traits, the vast majority of traits will be beyond any control and thus undergo changes of probably considerable extent. Hence, in a strict sense, dynamic "conservation" is difficult to realize for most of the characters of a population, but it may be feasible for a sufficiently small number of variable traits.

Yet, even traits readily amenable to observation may be difficult to conserve if they respond adaptively to certain environmental factors and if these factors vary in an unstable or unpredictable fashion. Insufficient control of the adaptively critical environmental factors may then result in maladaptation of the whole population, which would basically endanger the conservation success. Consequently, since unpredictably changing environments are the rule in natural as well as in artificial ecosystems, the enforcement of static principles to conservation measures *in situ* is inappropriate and even dangerous. Thus, if the term "conservation" is to be maintained in connection with dynamic processes, its meaning has to be extended to include as an essential constituent the preservation of adaptability of populations. The capacity of a population to adapt to variable environmental conditions is in turn determined by its genetic composition. Thus dynamic gene conservation is a measure for the preservation of adaptability.

In a recent paper dedicated to the development of general concepts of gene conservation and gene resources, ZIEHE et al. (1989) discussed the relative importance of static and dynamic gene conservation. The authors concluded that the biology, the environmental conditions, and the ecological position of many species give primacy to dynamic conservation; only in extreme situations, where a species or population is at the edge of extinction, may static methods provide the only *ad hoc* measures of conservation. Yet in order to. increase the chances for successful regeneration particularly of such gene resources, the transition to dynamic conservation ought to be realized as soon as possible. The reason is to be found in the increase of genetic variation after the allowance of sexual reproduction. A typical form of such a transition from static to dynamic conservation in forest trees is realized in "conservation orchards".

Building upon the above motivation, the present paper will focus on the elaboration of some basic characteristics of dynamic gene conservation that serve the preservation of adaptability. Moreover, the problem of appropriate population size will be discussed and demonstrated with the help of a model adapted to characteristics of reproductive systems of plants.

The Significance of Genetic Load

A genetic type is considered to contribute to the **genetic burden or load**, if the population fitness increases after subtraction of individuals of this type and their successful gametes. Thus, differential fitness of the genotypes in a population implies the presence of genetic load, but, at the same time, it is exactly this load which provides the basis for adaptive evolution. To arrive at a more detailed understanding of this view, consider the fact that individual fitness generally depends on both genotype and environment and that, therefore, the amount of genetic load may change if either

- for a population of given genetic structure its environmental conditions change, or if

- under given (stable) environmental conditions the genetic structure of a population changes.

Most adaptive processes will of course be characterized by changes in both genetic structures and environmental conditions, where the latter precede the former, so that, as a rule, adaptation lags behind environmental changes. However, in stable environments [*], evolution may proceed until a genetic equilibrium is reached. In asexually reproducing organisms this equilibrium will be freed from genetic load, since genetic fixation will be realized. With a sexual mode of reproduction, however, a genetic load can persist due to the possibility of polymorphic equilibria, as brought about by heterozygote advantage, for example. For certain systems of selection and mating this load is nevertheless minimized as compared with the non-equilibrium states.

Populations existing in environments which are changing in an unstable or unpredictable manner can generally be expected to carry higher genetic loads than those existing under stable conditions, and the amount of load is likely to increase with the speed and extent of such changes. This is a direct consequence of the above-mentioned fact that adaptation and thus the increase in frequency of the genotypes with the presently largest fitnesses always lags behind the environmental changes. In other words, the proportion of genotypes with suboptimal fitnesses under the respective environmental conditions will permanently be kept at a high level.

So far, the present considerations illustrate the conflicting roles played by genetic load in that it decreases population fitness on the one hand but presents itself as a necessary investment in a future of changing environments. This suggests that a compromise between the two roles should be realized in naturally reproducing populations.

---

[*] When applied to environments, the term "stable" is always assumed to refer to conditions that either do not change with time or change in a cyclically recurrent manner.

In fact, load is a concept defined in *relative* terms for genetic types within a given population, and it is usually quantified by the deviation of the maximum fitness among all genotypes from the population fitness, divided by this maximum. Hence, the same amount of genetic load can be realized in populations with high and with low average fitness if the maximum genotypic fitness is proportional to the population fitness. Since its (average) fitness exclusively decides on the survival of a population, genetic load appears to play no important role in this case. However, under stable environments such proportionality cannot occur, and it requires highly restrictive conditions under unstable environments. Moreover, fitness cannot increase indefinitely, so that increasing genetic load must eventually reduce population fitness to an extent that endangers survival.

Clearly, for given genotypic fitnesses, the population fitness would increase and genetic load would decrease as the fittest genotype would become more frequent. This situation gains relevance if environmental conditions remain stable over a long period of time. Now envisage a change in environmental conditions favouring a genotype that was previously inferior in fitness. This genotype would initially have a low frequency if the previously favoured genotype had reached a high frequency. Thus, after the change, the population fitness would be reduced drastically. To prevent such reduction it is therefore required that, under stable environmental conditions, the fittest genotypes do not become too frequent. This in turn would imply a sizable genetic load and leads to

Conclusion 1: The capacity to maintain sizable genetic loads under stable environmental conditions serves the preservation of adaptability.

Genetic Load and genetic variation

As is implicit in the above explanations, genetic load requires genetic variation, but the amounts of the two need not be correlated. The same load can be realized by any arbitrary number of more than one genetic type. This raises the question as to the possibility that genetic load and genetic variation have different effects on the preservation of adaptability.

In general, with an increasing range of environmental conditions to be survived, the number of different genotypes adapted to these conditions can also be expected to increase. Concerning the genetic load present in a population at any time, however, it makes a difference whether this range of heterogeneous conditions is distributed over space or over time. Spatially heterogeneous but temporally homogeneous (stable) environments combined with restricted gene flow allow for the evolution of local genetic differentiation by local adaptation. Therefore, large genetic variation can be maintained without considerable genetic load since the chances for genotypes to exist under more or less optimal conditions increase. However, extensive gene flow prevents such local adaptation, so that the actually existing spatial heterogeneity is experienced by the population as if the environment were effectively homogeneous. Hence, having the possibilities for locally differential adaptation in mind, any evaluation of spatial environmental heterogeneity should be carried out in terms of the "effective" heterogeneity by taking into consideration the equalizing effects of gene flow.

A completely different situation arises for the other extreme of (effectively) spatially homogeneous and temporally heterogeneous (unstable) environments. A considerable number of suboptimal genotypes must then exist at appropriate frequencies in each generation as a reservoir for the numerous environmental demands in the following generations. Hence, in this case, the persistence of a sizable genetic load will be inevitable if the genotypes vary strongly in fitness over the environments. The more genotypes there are that respond differently to a multitude of environments realized at different times, the more difficult it becomes to maintain an average fitness that allows survival of the population.

The rationale behind this statement stems from the concept of the norm of reaction of a genotype. Recall that the norm of reaction for a specified trait (such as fitness) is defined by the responses of a genotype to various environmental conditions. In fact, the above difficulty of maintaining a sufficient population fitness arises only if each genotype has a very narrow norm of reaction (with respect to fitness) that does not overlap with that of any other genotype. Consequently, there are two situations in which the problem that high genetic multiplicity causes excessive genetic load can be avoided:

(i)   the genotypes are characterized by broad norms of reaction with small overlapping zones of environmental conditions in which fitness is suboptimal, or

(ii) the norms of reaction are narrower but show substantial overlap in zones of optimal fitness.

In case (i) the whole spectrum of environmental conditions can be covered by a comparatively small number of genotypes, so that each genotype can be represented in sufficient frequency, thereby lowering the genetic load. The same effect can be achieved in case (ii), since the overlap in the norms of reaction allows several genotypes to respond equally well to a single environmental condition. This provides the chance of increasing the cumulative frequency of genotypes with optimal fitness under the respective environmental conditions. In summary:

Conclusion 2: The danger of excessive genetic load can be expected to predominate in environments that are heterogeneous in time and "effectively" homogeneous in space. A reduction of load and an improvement of the conditions for the preservation of adaptability can be achieved in this case if either the fitness norms of reaction of the genotypes have distinct but broad optima, or if many genotypes exist with narrower but extensively overlapping optima in their norms of reaction.

The distinction between genotypes with broad and with narrow fitness optima corresponds, of course, to the well-known distinction between generalists and specialists on the species level. According to the general reasoning that flexible response of an individual to variable environmental conditions requires genetic flexibility, genotypes with broad norms of reaction (generalists) can be expected to show higher degrees of heterozygosity than do genotypes with narrow norms (specialists). However, taking into consideration the possibility that different alleles at a gene locus may be functionally similar or even equivalent, the "effective heterozygosity" (referring to alleles with distinct functional differences) may be considerably smaller than the heterozygosity estimated on the basis of simple allele differences. This may in some cases lead to an overestimation of the frequency of generalists. On the basis of fitness measurements, however, the existence of homozygous genotypes with homeostatic capacity comparable to that of heterozygous genotypes cannot be excluded. For example, phenotypic modifications caused by environmental changes (plasticity) may preserve fitness under these environmental conditions (thus acting homeostatically with respect to fitness) without necessitating heterozygosity.

On the other hand, functional similarity of genes plays an important role for the reduction of genetic load among specialists, since it is the prerequisite for the existence of genotypes with overlapping fitness optima. This does of course not rule out the occurrence of genes with more expressed functional differences, which are necessary for covering extensive environmental differences. Consequently, in sexually reproducing populations, both individuals with low and with high "effective heterozygosity" may occur, which would suggest the persistence of a mixture of specialists and generalists. Yet the proportion of each critically depends on the mating system realized in the population. For example, inbreeding or positive assortative mating will increase the share of homozygotes and thus of specialists. Specialism is therefore most likely to be found in populations or species practicing predominantly homotypic mating.

Principles of dynamic gene conservation

It has become clear in the preceding considerations that the preservation of adaptability requires mechanisms guarding against loss of genetic diversity. Yet the persistence of genetic diversity during the course of adaptational processes need not accord with all of the objectives of gene conservation. This problem may arise, for example, if environmental changes result in the adaptive substitution of a resident gene by a previously disadvantageous mutant. Even though a qualitative change is implied by this substitution, the quantity of genetic diversity (usually measured as the effective number of genetic types) may remain roughly the same. Hence, in this case, conservation of previously existing genetic types would have failed, while the conservation of genetic diversity would not have been impaired.

The last example gives rise to the formulation of an important principle of dynamic gene conservation which is in accordance with the priority that must be given to the preservation of the adaptability of a gene resource population simply to secure its persistence under uncontrollable environmental changes. If the realized average fitness indicates population growth, management measures which increase genetic load are meaningful provided they lead to a reduction in the number of adaptive substitutions or to even an increase in genetic diversity. Since the consequences of introducing new genetic types are difficult to predict, this opportunity for increasing the genetic diversity should be handled with care. In some cases it might therefore be less problematical to consider measures that increase diversity by equalizing the proportions among the existing genetic types, i.e. by increasing the evenness.

At least in the initial phase, such measures must be accompanied by constant observation of the rate of change of population density (or population fitness) in order to be able to relax the load pressure if the population density decreases too drastically. Moderate and temporary decreases in population density can be tolerated, since they are unlikely to result in major changes of the frequencies of the genetic types. However, this decline must come to a halt as soon as a population size is reached that endangers the conservation success by genetic drift (the problem of critical population sizes will be returned to later). In summary:

Conclusion 3: Avoidance of genetic substitutions and the maintenance or even increase in genetic diversity (preferably by increasing the evenness) are basic concerns of dynamic gene conservation. These aims may justify measures which raise the genetic load to a degree that implies moderate but only temporary declines in population density. Such measures must be relaxed at the latest when, due to small population size, drift effects may have a chance to reduce the genetic diversity.

Environmental and population design

Preservation of the adaptability of a resource population with minimum genetic load, maximum avoidance of genetic substitutions, and maximum genetic diversity and evenness would, of course, be the ultimate object of endeavour of all measures of dynamic gene conservation. The forces affecting these genetic components are highly complex and therefore difficult to identify and to control. Yet, the preceding considerations allow some general strategies for the design of a resource to be outlined.

The close association between the degree of environmental heterogeneity and genetic diversity was emphasized, and it was argued that temporal heterogeneity is likely to be accompanied by larger amounts of genetic load than is spatial heterogeneity. Therefore, a reduction of genetic load that avoids losses of genetic diversity is most likely to be achieved by replacing as much of the temporal environmental heterogeneity as possible by spatial heterogeneity. Thus the spatial heterogeneity would increase at the expense of the temporal, which, by definition, would stabilize the environmental variation. There are many environmental factors that can be considered in this context. Water supply, for example, may vary irregularly over time due to the exposition, A say, of a population. At another exposition, B say, the overall water supply may be nearly constant over time (including fixed cyclical changes), but the spatial differences could cover the same range of variation as is realized over time at exposition *A*. Hence, the temporal heterogeneity at *A* is replaced by spatial heterogeneity at *B*, so that exposition *B* would be preferable for the establishment or declaration of a gene resource. Similar examples can easily be named for biotic environmental factors.

Besides reducing genetic load, stable spatial heterogeneity has the advantage of allowing for regulation of the population density realized under the various environmental conditions. For example, if all environmental conditions are represented by the same number of individuals, this would help to realize another of the above-mentioned objectives, namely to increase the evenness in representation of the genetic types. However, it has of course to be taken into account that measures regulating population density may produce new environmental conditions via competition and may affect the mating system. Modifications of the mating system must in turn be expected to alter the proportions of the genotypes in the next generation, so that attempts at locally differential regulation of population density may not have the desired effect. If possible, it is thus preferable to equalize the spatial representation of environmental conditions.

Conclusion 4: A lowering of genetic load without loss of genetic diversity can be achieved by shifting the required environmental heterogeneity from its temporal to its spatial component. Equalization of the spatial representation of the environmental conditions may additionally increase the genetic evenness.

Another advantage resulting from the stabilizing (recall the definition of "stable" given earlier) effect of the above replacement procedure refers to the distinction between "specialist" and "generalist" adaptive norms of reaction. Under stable spatial environmental heterogeneity the proper distinction between these types of norms is probably less mandatory. However, since temporal environmental heterogeneity can never be excluded completely, the statements in Conclusion 2 are still relevant.

The significance of population size

So far the deliberations have concentrated on general adaptive aspects that ought to be considered in the declaration, establishment, and management of dynamically conserved gene resources. While these deliberations remain valid irrespective of the population sizes addressed, a more precise quantification of the risks of losing genetic variation must explicitly account for the possible effects of population size, among which random events resulting in genetic drift are the most prominent. Concerning genetically selective forces, which are part of the desired adaptational processes, genetic drift may weaken the efficiency of mechanisms protecting polymorphisms, or it may accelerate the process of gene substitution implied by directional selection.

Irrespective of the population size, selective gene substitution is difficult to guard against, and it probably even should not be done since it may indicate an adaptive process required to free the population from excessive genetic load. On the other hand, if the active selective forces do not entail genetic substitution, the risk of losing genetic variants would be greatest if selection were absent altogether, i.e. if the genetic variation were selectively neutral. Hence,

Conclusion 5: The search for desirable population sizes should be oriented at the extent of genetic drift within selectively neutral polymorphisms.

This recommendation is valid irrespective of the conflicting views about the actual extent of selectively neutral genetic variation, since it refers to the worst possible situation that can arise with dynamic gene conservation and which should therefore be the point of orientation. Moreover, a certain portion of the genetic variation can be non-adaptive under some but not under other environmental conditions, which would expose this portion to the effects of drift if the non-adaptive environment persists over a sufficient length of time. In any event, the motivation to maintain population sizes reducing drift within sets of selectively equivalent genes stems from the expectation that these genes may eventually become adaptive under changed environmental conditions (see e.g. STEBBINS and HARTL, 1988). Otherwise genetic drift could hardly be a matter of concern.

Measurement of genetic variation

Any quantification of genetic drift must of course rely on measurements of genetic variation, and the rate of change of these measurements in the course of time then specifies the extent of genetic drift. Even though in almost all studies on this matter heterozygosity serves as the basic measure of genetic variation, this choice is problematical since heterozygosity may vary from 0 to 100% as long as at least two alleles exist at a locus and if the most frequent allele does not exceed 50%. Hence, heterozygosity provides almost no information on the number and frequency of genes existing in a population. The only exception is provided by Hardy-Weinberg proportions, but the realization of such genotypic frequencies requires highly specific stipulations. In fact, the main concern of drift studies lies in observing the conditions for loss of genes rather than genotypes, simply because genes are the raw material from which genotypes can be formed in various ways under the action of a multitude of recombination and mating systems.

Consequently, as applied to studies of genetic drift, measures of genetic variation should reflect the variation contained in the gene pool *. Essentially, there are two methods of measuring the variation of a population's gene pool: one reflects the relative and the other the absolute amount of genic variation. The former is known as "gene pool differentiation" and the latter as "gene pool diversity". For a population of size N and a single diploid gene

---

* For a specified collection of gene loci the gene pool of a population is defined to be the set of all individual genes present at these loci in the members of the population.

locus with relative frequency q_i of the i-th allele, the gene pool differentiation is given by (see GREGORIUS 1987)

$$\delta_T = \frac{2 \cdot N}{2 \cdot N - 1} \cdot \left[1 - \sum_i q_i^2 \right]$$

The gene pool diversity is in this case usually specified as the "effective number of alleles" (CROW and KIMURA 1970, p. 323) which is equivalent to the "differentiation effective number of alleles" (GREGORIUS 1987):

$$v = \left[\sum_i q_i^2 \right]^{-1}$$

Note that δ_T and n depend on each other according to the generally valid equation

$$\delta_T = \frac{2 \cdot N}{2 \cdot N - 1} \cdot \left[1 - \frac{1}{v} \right] \tag{1}$$

The ranges of variation of δ_T and n are [0,1] and [1,2N], respectively, where $\delta_T = 0$ and n = 1 characterize the state of genetic fixation for a single allele, while $\delta_T = 1$ and n = 2N implies that all of the 2N individual alleles differ in state. When the gene pool is specified for more than one gene locus, the arithmetic and the harmonic mean of the single locus measurements have to be taken for δ_T and n in order to obtain the gene pool differentiation and the gene pool diversity, respectively (GREGORIUS 1987).

To decide on which of the two measures is more suited to which purpose, it is helpful to recall that population size delimits the maximum possible number of alleles per locus. Hence, a reduction in the number of alleles and thus of genetic diversity can be obtained by merely decreasing the population size. Therefore, the gene pool diversity n is affected by the population size, which is not the case for the gene pool differentiation δ_T. For a given size of the resource population, δ_T rather measures the degree to which the capacity of this resource to harbour genetic variation is exhausted. In other words, δ_T represents the state of genetic saturation of the resource while v represents its (absolute) genetic richness.

When establishing or declaring a gene resource, its genetic richness is of primary interest, and the larger v is, the more individuals may be required to cover the diversity. The problem is then to determine a sufficiently large population size guarding the diversity against intolerable losses due to drift. As is seen from equation (1), the genetic saturation δ_T decreases for given diversity v with increasing population size N. However, this tendency levels out very quickly, so that further increases of N have practically no observable effects. In the vast majority of realistic situations, δ_T and v are therefore essentially in one-to-one correspondence, so that studies of drift effects can refer to any of the two measures.

However, particularly in large populations, where δ_T and v have to be estimated from random samples, the inestimability of v reveals δ_T as the definitely superior measure of gene pool variation. Moreover, since the size of a resource population limits its genetic diversity *a priori*, the most important question concerns the effect of population size on the maintenance of certain levels of genetic saturation.

Conclusion 6: The determination of population sizes guarding from excessive genetic drift should be based on the rate of change in gene pool differentiation (genetic saturation) δ_T.

Effects of the reproductive system

Obviously, population size need not be the sole determinant of the extent of genetic drift. If in each generation on the average only 80% of the members of a population are fertile, the "drift effective" population size is reduced by at least this percentage. However, the effects of other components of the reproductive system, such as sexual or mating system, are less obvious. The earliest result on effects of the sexual system is due to WRIGHT (1931), who showed that unequal sex ratios in dioecious populations accelerate the speed of genetic drift. This approach can be generalized to include arbitrary fertilities and mating relations among the individuals, so that preferential mating among neighbouring individuals, self-fertilization, simultaneous occurrence of unisexual and bisexual individuals, environmentally induced fertility variation, sexually asymmetric or asynchronous flowering, etc., appear as special cases (all of these forces acting, by definition, independently of the gene locus under consideration). For organisms producing large numbers of zygotes per parent (such as plants), it was demonstrated by GREGORIUS (1986) that each of these variables affects genetic drift only as far as it has a bearing on any of the two parameters: (*i*) reproductively effective number n_r of individuals (or the reproductively effective population size) and (*ii*) the proportion s of self-fertilizations among all offspring of the population. The reproductively effective population size n_r is defined in complete analogy with the genetic diversity $1/\Sigma q_i^2$ (which is also an effective number), where now q_i is the expected proportion of successful gametes of the *i*-th individual among all successful gametes of the population. Herewith a gamete is termed successful if it entered a zygote.

Denoting by $\delta_T(t)$ the expected genetic differentiation of the population at generation *t*, the rate of change in genetic differentiation in two successive generations can be specified by the multiplication rate

$$d := \frac{\delta_T(t)}{\delta_T(t-1)}$$

According to the previous explanations, *d* measures the speed of genetic drift (actually *d* is inversely related to the speed of drift). For the purpose of dynamic gene conservation the long term consequences of a presently realized situation is of primary concern, so that the value that d assumes after a sufficiently large number of generations with constant reproductively effective number n_r and proportion s of self-fertilization is required. For a single gene locus it can be proven that, in the course of the generations, d approaches a limiting value $\hat{d}$ given by

$$\hat{d} = \frac{1}{2} \cdot \left(1 - n_r^{-1}\right) + \frac{1}{4} \cdot s + \frac{1}{2} \cdot \sqrt{1 - s + \left[\frac{1}{2} \cdot s - n_r^{-1}\right]^2} \qquad (2)$$

(see Appendix). Besides quantifying the asymptotic speed of drift, this result provides us with the following basic insights:

Conclusion 7: For a single selectively neutral gene locus, the asymptotic rate of genetic drift as measured by the multiplication rate $\hat{d}$ of expected genetic differentiation obeys $\frac{1}{2} \leq \hat{d} < 1$ and is completely specified by the reproductively effective population size n_r and the population selfing proportion s. Moreover, $\hat{d}$ decreases (i.e. the speed of drift increases) with increasing s and decreasing n_r.

Even though these statements were derived for separated generations, their principles can be expected to extend to overlapping modes of reproduction provided the generation time is adjusted accordingly. It is in any case of considerable interest to know the extent to which n_r and s affect the reduction $\hat{d}$ in differentiation per generation. This knowledge allows practical decisions to be made on the size and structure of a gene resource on the basis of the proportion of genetic differentiation that can be expected to be lost after a certain number of generations. For example, if for any reasons it is tolerable to lose 5% of the original genetic differentiation within the next 10 generations, $\hat{d}$ ought to be not less than $(1 - .05)^{1/10} \approx 0.99488$. Given $\hat{d}$ as determined by this criterion, the problem then consists in finding the population size and the reproductive characteristics yielding $\hat{d}$. In fact, this statement of the problem probably reflects the practical aims of dynamic gene conservation most precisely.

Yet, according to equation (2) $\hat{d}$ depends on the reproductively effective rather than on the actual population size, where the first of course includes the effects of and never exceeds the last. Hence, the immediately relevant information is provided by n_r as a function of $\hat{d}$ and s. Solving equation (2) for n_r yields the desired result

$$n_r = \left[\frac{\frac{1}{4}.(1-s)}{\hat{d} - \frac{1}{2}} \cdot \hat{d} + \frac{1}{2} \cdot (1+s) \right]^{-1} \tag{3a}$$

In this equation $\hat{d}$ is in turn determined by the expected proportion u of the original genetic differentiation that is lost after t generations, so that

$$\hat{d} = (1-u)^{1/t} \tag{3b}$$

Both equations (3a) and (3b) together specify the reproductively effective population size required to limit the loss of genetic differentiation after t generations of drift to a fraction u.

Conclusion 8: The determination of the size of a resource population should be based on the expected fraction of the original genetic differentiation, the loss of which by drift is tolerable over a specified number of generations. The size of the resource should be stated in terms of the reproductively effective population size.

The use of reproductively effective in place of actual population sizes is mandatory, since many individuals either do not survive to the reproductive age or participate to varying degrees in the mating process. Such individual `redundance' may vary in a species-specific manner and has to be considered when evaluating the size of a resource population.

The effect of self-fertilization

It remains to demonstrate the effect of self-fertilization on genetic drift. That increasing amounts of self-fertilization increase the speed of drift by decreasing the multiplication rate $\hat{d}$ was already emphasized in Conclusion 7. However, since with growing reproductively effective population size $\hat{d}$ approaches a value of 1 irrespective of the value of s, one might conclude that the effect of s is negligible for sufficiently large n_r. Yet, this conclusion would be misleading, as can be seen by computing n_r from equation (3a) for the extreme situations $s = 0$ and $s = 1$. Denoting the corresponding n_r-values by $n_r^{(0)}$ and $n_r^{(1)}$, it turns out that

$$\frac{n_r^{(0)}}{n_r^{(1)}} = 1 - \frac{1}{2 \cdot \hat{d}}$$

Consequently, $n_r^{(1)} > 2 \cdot n_r^{(0)}$ for any value of $\hat{d}$, i.e. the reproductively effective size of a non-self-fertilizing species needs to be more than twice that of a completely self-fertilizing species in order to maintain the same speed of drift. Since in most of the relevant cases the tolerance parameters u and t in equation (3b) yield $\hat{d}$ values very close to 1, $n_r^{(1)} \approx 2 \cdot n_r^{(0)}$ is a good approximation. Table 1 lists reproductively effective population sizes for both extreme proportions of self-fertilization and for various proportions u and generations t.

Tab. 1: Reproductively effective population size n_r required to lose not more than u %, on the average, of the original genetic differentiation within t generations of drift. The left and right numbers refer to n_r in the absence of and for complete self-fertilization, respectively.

$u(\%)$	Generations t					
	5	10	15	20	25	30
1	248 - 498	497 - 995	746 - 1493	995 - 1990	1243 - 2488	1492 - 2985
2	123 - 248	247 - 495	371 - 743	495 - 990	618 - 1238	742 - 1485
3	82 - 165	164 - 329	246 - 493	328 - 657	410 - 821	492 - 985
4	61 - 123	122 - 245	183 - 368	245 - 490	306 - 613	367 - 735
5	48 - 98	97 - 195	146 - 293	195 - 390	243 - 488	292 - 585
6	40 - 81	81 - 162	121 - 243	161 - 324	202 - 405	242 - 485
7	34 - 69	69 - 138	103 - 207	138 - 276	172 - 345	206 - 414
8	30 - 60	60 - 120	90 - 180	120 - 240	150 - 300	180 - 360
9	26 - 54	53 - 107	79 - 160	106 - 213	132 - 266	159 - 319
10	23 - 48	47 - 95	71 - 143	95 - 190	118 - 238	142 - 285

According to the preceding explanations, the numbers in Table 1 constitute lower bounds, since the actual population size required to yield the given reproductively effective sizes may exceed the latter considerably, and, in addition, the derivations are based on a single gene locus. Concerning the number of selectively neutral gene loci to be accounted for, at least two points of view are of interest. One derives from the fact that the ultimate rate of decay of genetic differentiation does not depend on the initially realized amount of genetic variation and, therefore, is representative of any arbitrarily variable selectively neutral gene locus. Even though it is oriented at a single locus, this view can be extended to cover multiple loci, since the probability of losing a certain amount of genetic differentiation at two loci with 5 alleles at each, say, is likely to be smaller than for a single locus with 10 alleles. Hence, population sizes accounting for drift effects at a single gene locus might as well be applied to multiple loci. Yet, this line of reasoning can obviously not be extended to larger numbers of loci, since the numbers of different genes across all of these loci could readily exceed twice the population size, which would render the one-locus analogue unrealistic.

The other point of view proceeds from the conception that, as a rule, environmental changes turning previously selectively neutral into adaptive gene loci require co-adapted gene complexes extending over several loci. Under this premise it is difficult to justify any upper limit that could possibly be set to the number of gene loci to be considered for the decay of gene pool differentiation. If there is a solution to this problem at all, it is likely to be provided by a proof that the rate of decay in gene pool differentiation approaches a non-trivial upper bound with increasing numbers of gene loci. This would then account for both of the above points of view.

Concluding remarks

Dynamic gene conservation as presented here of course also contributes to species protection, in that it explicitly considers the natural conditions for persistence of populations which are *per se* representatives of species. In general, population protection includes two basic types of measures between which primacy is commonly given to the securing of the (species-specific) demographic characteristics (such as habitat requirements, population density, subpopulation structure, age class structure, timing of sexual maturity, etc.) required for survival and reproduction (see e.g. LANDE 1988). While this type of measure is clearly indispensable, it cannot be viewed independently from the second type, which is concerned with provision of the genetic variation required for adaptation and preservation of adaptability. In fact, demographic characteristics may appear as environmental conditions to which the genotypes must be adapted, or they may represent traits whose expression depends on the genotype.

Thus, the objectives and methods of population protection and dynamic gene conservation appear to be largely identical, with the important difference, however, that a major concern of dynamic gene conservation may consist in describing the conditions under which amounts of genetic multiplicity can be maintained in excess of those required for the preservation of adaptability. As became apparent in the deliberations leading to Conclusions 1 to 4, this critically depends on the possibility of estimating genetic loads implied by the association of the genetic structure of a population with its environmental conditions, and on the possibility of judging the degree to which these loads can be tolerated without endangering adaptability. While it may be difficult to develop appropriate methods of estimation, general guidelines such as those addressed in Conclusions 2, 3, and 4 can nevertheless be inferred.

Environments serving purposes of dynamic gene conservation cannot be specified without consideration of the basic demographic characteristics of the species concerned. Therefore, as is the case with population and species protection, the supporting (biotic and abiotic) environment of the gene resource must be included in any measure of conservation. This in turn necessitates ecosystem protection. Though evident in conclusion, the awareness that, even for practical reasons, gene conservation is very unlikely to succeed in disturbed ecosystems has increased only recently (see e.g. BRUSH 1989).

Zusammenfassung

Als unabdingbares Ziel einer Gen-Konservierung wird der Erhalt der Anpassungsfähigkeit einer Ressourcenpopulation herausgestellt. Daher müssen Methoden einer dynamischen Konservierung Priorität erhalten. Zur Anpassungsfähigkeit bedarf es der Existenz genetischer Variation, welche unter den jeweils vorherrschenden Umweltbedingungen entweder adaptiv unterlegen oder neutral ist. Hiervon ausgehend wird die Bedeutung der "genetischen Last" und der Umweltheterogenität für die Aufrechterhaltung genetischer Variation in den Ressourcenpopulationen diskutiert und einige Hauptprinzipien, die bei der Gen-Konservierung in Betracht gezogen werden müssen, werden herausgestellt. Zur Festlegung der Größe von Ressourcenpopulationen wird als Kriterium vorgeschlagen, sich an dem Verlust genetischer Variation aufgrund genetischer Drift zu orientieren, der über eine bestimmte Anzahl von Generationen toleriert werden kann. Es wird ein entsprechendes Modell vorgestellt, und die Populationsgrößen, welche sich für verschiedene Reproduktionssysteme ergeben, werden damit berechnet. Die Beziehungen zwischen einer dynamischen Gen-Konservierung und dem Schutz von Populationen (Arten) werden hervorgehoben und die Notwendigkeit des Schutzes von Ökosystemen wird dargelegt.

References

BRUSH S.B. (1989): Rethinking crop genetic resource conservation. Conservation Biology 3(1): 19-29

CROW J.F., M. KIMURA (1970): An Introduction to Population Genetics Theory. Harper & Row, New York, Evanston, London

FORD-LLOYD B., M. JACKSON (1986): Plant Genetic Resources: An Introduction to their Conservation and Use. Edward Arnold Publ.

GREGORIUS H.-R. (1986): The change in heterozygosity and the inbreeding effective size. Math. Biosci. 79: 25-44

GREGORIUS H.-R. (1987): The relationship between the concepts of genetic diversity and differentiation. Theor. Appl. Genet. 74: 397-401

LANDE R. (1988): Genetics and demography in biological conservation. Science 241: 1455-1460

SCHONEWALD-COX C.M., S.M. CHAMBERS, B. MACBRYDE, L. THOMAS (1983): Genetics and Conservation: A Reference for Managing Wild Animal and Plant Populations. The Benjamin/Cummings Publishing Company

STEBBINS G.L., D.L. HARTL (1988): Comparative evolution: Latent potentials for anagenetic advance. Proc. Natl. Acad. Sci. USA 85: 5141-5145

WRIGHT S. (1931): Evolution in Mendelian populations. Genetics 16: 97-159

ZIEHE M., H.-R. GREGORIUS, H. GLOCK, H.H. HATTEMER, S. HERZOG (1989): Gene resources and gene conservation in forest trees: General concepts. In F. SCHOLZ, H.-R. GREGORIUS, D. RUDIN (eds.) Genetic Effects of Air Pollutants in Forest Tree Populations. Springer-Verlag

Appendix

To derive equation (2) the model of GREGORIUS (1986) can be restated in terms of δ_T. In this model the effects of drift are considered in terms of the variables f and g, where

$f_t :=$ probability that at a given (diploid) gene locus the two genes of an individual in generation t are identical in state, and

$g_t :=$ probability that at a given (diploid) gene locus the two genes drawn at random from two different individuals in generation t are identical in state.

The dynamics of the model are analyzed in terms of the rates $h_t := (1-f_t)/(1-f_{t-1})$ and $z_t := (1-g_t)/(1-f_t)$, and it is shown that for constant parameters of the reproductive system with $s < 1$ these rates converge.

Denoting by N_t the actual population size in generation t, it follows from the definition of δ_T that

$$\delta_T(t) = \frac{2(N_t - 1)(1 - g_t) + (1 - f_t)}{2 N_t - 1}$$

Thus

$$d := \frac{\delta_T(t)}{\delta_T(t-1)} = \frac{2N_{t-1} - 1}{2N_t - 1} \cdot \frac{2(N_t - 1) z_t + 1}{2(N_{t-1} - 1) z_{t-1} + 1} \cdot h_t$$

Consequently, for constant population size $N_t \equiv N$ one obtains $\hat{d} = \hat{h}$ at the limit. $\hat{h}$ is given in the above-cited paper, where the present notations s and n_r correspond to R and Q^{-1}, respectively, in this paper.

Population Extinction by Mutational Load and Demographic Stochasticity

W. Gabriel [1], **R. Bürger** [2] **and M. Lynch** [3]

[1] Dept. of Physiological Ecology, Max Planck Institute for Limnology, Postfach 165,
   D-2320 Plön, Germany
[2] Institut für Mathematik der Universität Wien, Strudlhofgasse 4, A-1090 Wien, Austria
[3] Dept. of Biology, University of Oregon, Eugene, Oregon 97403, USA

Abstract

Genetic aspects are important in the evaluation of the risk of extinction for small populations. Using estimates of rates and effects of slightly deleterious mutations, we calculate the mean time to extinction under the joint action of mutation load and density-dependent stochastic population regulation. Accumulation of mutations diminishes the individual survival probability, which leads to a reduction in population size. This, in turn, progressively facilitates the fixation of future deleterious mutations by random genetic drift. This synergistic interaction has been called the mutational melt-down.

In asexual populations, the probability of extinction increases as the mutational effect increases and as actual population size decreases. As reference points for sexual populations, we present the expected extinction times without mutational load but with stochastic fecundity and sex-ratio under a logistic population regulation. Selection and recombination does not prevent mutational melt-down in small sexual populations; slightly deleterious mutations reduce the mean time to extinction by several orders of magnitude. Stochastic fecundity is a minor direct source of extinction in sexual populations, but it leads to temporary reductions in effective population size, which increases the risk of extinction due to stochastic variations of the sex-ratio.

Introduction

Several well known risks for population extinction increase drastically with decreasing population size. Biotic and abiotic fluctuations of the environment or purely stochastic variation of demographic parameters (like birth and death rates, carrying capacities, and sex-ratio) can reduce a population to a level at which the probability of extinction is high. The smaller the number of individuals is, the more severe are genetic problems such as inbreeding depression, loss of adaptive variation by random drift, and reduction of fitness due to fixation of deleterious mutations. Theoretical aspects of the dynamics of populations in variable environments have received considerable attention (RICHTER-DYN & GOEL 1972; FELDMAN & ROUGHGARDEN 1975; HANSON & TUCKWELL 1978; MAY 1981; HOPPENSTEADt 1982; NISBET & GURNEY 1982; WRIGHT & HUBBELl 1983, MODE 1985; EWENS et al. 1987, GOODMAN 1987, SOULÉ 1987). However, there have been few attempts to incorporate explicit genetic details into extinction models. In the following, we give a review of recent work on the interaction of deleterious mutations, population dynamics and extinction.

In principle, any deleterious mutation can reduce the size of a population. The probability of fixation of such mutations by random genetic drift increases with declining population size, leading to a synergistic interaction which LYNCH & GABRIEL (1990) called a "mutational melt-down". This process is especially relevant for slightly deleterious mutations in small asexual populations. However, because recombination facilitates selection against bad mutations, it is believed that genetic risks are small compared to risks by demographic and environmental influences (LANDE 1988). This opinion will be questioned later on in this paper when we demonstrate the possibility of mutational melt-down in small sexually reproducing species.

Estimates for Mutation Rates and Selection Coeffiecients

The effectiveness of the mutational melt-down depends critically on the rate and effects of mutations. We do not consider lethal recessive mutations since we know from earlier simulations that their influence is small compared with mildly deleterious mutations. Many estimates of mutation rates and mutational effects are available for *Drosophila melanogaster*. According to the review of CROW & SIMMONS (1983) each animal incurs an average of 0.6 new non-lethal mutations, each of which reduces the viability by ≈2.5%. Such mutations appear to be approximately additive within loci (MUKAI 1979). If mutations on different loci act independently then the fitness reductions from each locus can be treated as multiplicative. We will adopt this assumption, although data are not precise enough to rule out other interpretations.

Therefore, letting W' be the fitness of an organism that carries slightly deleterious mutations at n loci and W_0 be the fitness of an organism without these mutations, we have

$$W' \quad = \quad W_0 \, (1 - s_1)(1 - s_2) \ldots (1 - s_n) \, , \tag{1}$$

where s_i is the selection coefficient at locus i. Assuming that each new mutation occurs at a different locus and that the selection coefficient is the same for all these loci, one gets

$$W'/W_0 = (1 - s)^n \approx (1 - s)^{\mu t} \tag{2}$$

$$\approx 1 - \mu s t \quad \text{for small } \mu s t$$

with t as time and with μ as the zygotic mutation rate per time unit and with μt as the expected number of mutations. With these assumptions, LYNCH & GABRIEL (1990) estimated the mutation load for organisms other than *Drosophila* using data from experiments designed for other purposes. For eukaryotes their estimated boundaries for the mutation load μs are

$$0.0002 < \mu s < 0.02. \tag{3}$$

Within the range of realistic parameters, the melt-down process is determined mainly by the product μs so μ and s need not be known separately.

Deleterious Mutations in Asexual Populations

Although asexual species are often considered to be evolutionary dead-ends, polygenic mutation provides an evolutionary potential which is sufficient for considerable phenotypic evolution (LYNCH & GABRIEL, 1983). Nevertheless, a severe handicap for asexual species is the accumulation of unconditionally deleterious mutations. If, by chance, the genotype with the fewest deleterious mutations does not contribute to the successful offspring in the next generation, then this genotype is removed from the population forever. Eventually, the second best genotype will have the same fate of being lost from the population - and so on. MULLER (1964) first noticed this process, which has become known as "MULLER's ratchet". The velocity at which the ratchet turns, depends on the population size, mutation rate, and selection coefficient.

Earlier theory and simulations of the ratchet (e.g. MAYNARD SMITH 1978, BELL 1988) kept the (effective) population size constant. This led to the prediction that the ratchet is less effective for higher selection coefficients; i.e., the average time for the ratchet to make one turn increases with increasing s. However, by making the population size dependent on mutation load, LYNCH & GABRIEL (1990) found that the mean extinction time declines as mutations become more deleterious. This means that the lower speed of the ratchet under stronger selection (higher s values) is more than compensated for by the greater reduction in survivorship per turn of the ratchet. A synergistic interaction between mutation load and random genetic drift is responsible for this process. As deleterious mutations reduce the number of surviving offspring, random genetic drift becomes of greater importance and facilitates the fixation of further deleterious mutations. This "mutational melt-down" eventually leads to population extinction.

Figure 1 gives estimates of extinction time for asexual populations under a low mutational load of μs = 0.0002 for various intrinsic growth rates r. These results are obtained by a model with very simple population density regulation and non-overlapping generations. Each member of the population can produce $R = e^r$ offspring on average but offspring numbers are Poisson distributed around this expectation. The population size is restricted to K individuals (= carrying capacity). If the total number of offspring exceeds the carrying capacity, K individuals are drawn randomly to start the next generation. The accumulated mutations in each offspring determine its probability of survival until reproduction. For small μs, the combined effect of these random processes can be approximated by a deterministic solution which can easily be calculated by iterations (for details, see LYNCH & GABRIEL 1990).

The results of this conservatively simple model, which ignores all environmental sources of mortality, suggest that the survival times of clonal lineages of a carrying capacity $K < 10^7$ are unlikely to exceed 10^4 generations. This view is consistent with molecular data that suggest that most parthenogenetic animals are phylogenetically young (BELL 1982, LYNCH 1984). It is in contradiction to the existence of a few very old obligate parthenogenetic groups (e.g. bdelloid rotifers). Assuming errors have not been made in the identification of the breeding system of such groups, their escape from mutational melt-down may be a consequence of a high incidence of compensatory mutation. LYNCH & GABRIEL (1990) modelled the melt-down process using a distribution of mutational effects with a constant mean. They found that the longevity of asexual lineages can be enhanced dramatically if the variance in s becomes large enough so that some mutations are beneficial.

It should be mentioned here that the above statements on MULLER's ratchet for the case of constant mutational effect rely on the assumption of unconditionally deleterious mutations. This means that mutational effects are independent of the actual genetic background. An alternative treatment involves the influence of mutations on quantitative traits (GABRIEL & WAGNER 1988; WAGNER & GABRIEL 1990). In this case, fitness is determined by several quantitative traits, each trait of which is controlled by many loci. A single pleiotropic mutation can affect each trait simultaneously. Since polygenic mutations can increase or decrease the value of a trait, deleterious mutations for one trait or locus can be compensated for by advantageous mutations for others, even when the average effect of

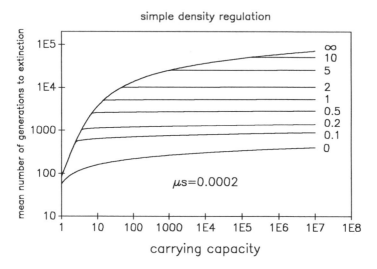

Fig. 1:   Survival of parthenogenetic populations under mutational load ($\mu s = 0.0002$) and simple density- dependent population regulation (see text). The mean number of generations to extinction is calculated as a function of carrying capacity K for various intrinsic growth rates r (in units per generation as indicated by the numbers in the graph).

each mutation is a reduction of fitness. This model is a quantitative genetic analog to MULLER's ratchet. Simulations and mathematical analyses show that the compensatory mutations inherent under this scenario are as effective as recombination in halting the decline of fitness caused by MULLER's ratchet.

A Simple Logistic Model with Stochasticity in Sex-Ratio and Fecundity

The simple population regulation described in the previous section is not very realistic but sufficient for initial insight into the interplay between genetics and population dynamics. In the following we will use a model with a logistic type of population growth. This also can be parameterized by the carrying capacity K and the growth rate r. (K is the population size, to which an undisturbed population would converge during time; r is equivalent to the intrinsic rate of exponential growth at small population size.) These parameters are usually used in differential-equation models where time is a continuous variable. An analogous difference-equation description for discrete generations is

$$N_{(t+1)} = N_{(t)} \ e^r \ [1 + N_{(t)}(e^r - 1)/K]^{-1}. \qquad (4)$$

With this equation there are no periodic orbits or chaotic behavior (for details see MAY 1981). Instead, convergence to K occurs in a monotonic way.

We use equation (4) to calculate the expected number of offspring per female. For asexual populations mean family size would be just $N_{(t+1)}/N_{(t)}$. For sexual populations we assume an expected ratio of females to males of 1:1. In order to achieve the same population regulation as in the asexual case, each female has to produce $2N_{(t+1)}/N_{(t)}$ offspring on average. The actual number of newborns is drawn from a Poisson distribution. Each female randomly chooses a mate, and the offspring's genome is constructed by free recombination. For this purpose, each mutation and its locus is stored. Every new mutation is assumed to occur at a new locus.

Thus, the model we now consider involves logistic growth but with stochasticity in fecundity and in the sex-ratio. There are two potential causes of extinction: either the number of surviving offspring is zero by chance, or mating is impossible because there are only males or only females left.

The Risk of Extinction from Random Variation in Fecundity and Sex-Ratio without Mutational Load

Before we study the combined effect of demographic stochasticity and mutational load, we analyze the risk of extinction without mutations. To get reference points for evaluating the relative importance of the underlying processes, we first look at the consequences of sex-ratio fluctuation when the population size is kept at the carrying capacity. Then, neglecting the sex-ratio fluctuations, we study the effect of stochastic fecundity in a monoecious population with density- dependent offspring number. Finally, we calculate the expected extinction times under the simultaneous operation of both stochastic processes for logistic population regulation.

The extinction probabilities due to sex-ratio fluctuations can easily be calculated for constant population sizes. (Further details are given elsewhere; see GABRIEL and BÜRGER, submitted). For a population of size K the mean time to extinction due to sex-ratio variation is

$$t_E = 2^{K-1} + 1,$$ (5)

given an expected sex-ratio of 1:1. The extinction times are geometrically distributed so that the standard deviation is equal to the expectation.

Without sex but under stochastic fecundity and logistic population regulation, there exists no simple analytic expression for the distribution of extinction times. For the combined process, it is even more hopeless to find analytical solutions other than approximations for special cases. Therefore, to get reliable results, we applied two

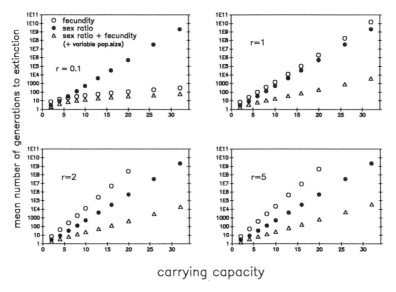

Fig. 2:   Extinction times due to stochastic fecundity and/or stochastic sex-ratio depending on carrying capacity K and intrinsic growth rate r (which is measured in units of inverse generation time; therefore, r = 1 implies 2.72 offspring and r = 5 leads to 148 offspring on average.). The values for stochastic sex-ratio (closed circles) are calculated for constant population size (= K). For stochastic fecundity (open circles), the expected number of offspring is regulated by logistic growth but the number of offspring produced by the individual females fluctuates around this expectation value according to a Poisson distribution. The populations start in the first generations with K individuals. The results of the combined processes of stochastic sex-ratio and stochastic fecundity are given by the triangles.

independent methods: a) straight-forward Monte Carlo simulations, and b) description of the processes by Markov chains and numerical solution of the corresponding (quite large) systems of equations. For details see GABRIEL and BÜRGER (submitted). We obtained identical results from both methods so that we can be sure that there are no errors in the program for the Monte Carlo simulation and that we did not run into numerical problems in the solution of the equations of the Markov chain model.

Figure 2 compares the extinction times under the influence of stochastic sex-ratio alone, stochastic fecundity alone, and under the combined action of both for various growth rates and carrying capacities, with the populations always starting at the carrying capacity (K). Only for small r is the probability of extinction due to sex-ratio variation larger than that due to stochastic fecundity. For increasing r the risk of extinction due to stochastic fecundity converges rapidly to an analytically calculable limit, e.g. mean time to extinction converges to e^K if only stochastic fecundity is considered (see GABRIEL and BÜRGER, submitted).

There is of course no linear interaction between the pure sex-ratio and fecundity risks because in the combined process the risk due to sex-ratio is not determined by the carrying capacity K but by the actual population size in each generation. An analysis of causes for extinction shows that the probability of population extinction due to non-surviving offspring becomes very small as the carrying capacity increases. Figure 3 shows the extinction risks due to the combined stochasticity of fecundity and sex-ratio depending on K for various r values.

One should keep in mind that the plotted numbers are expected times to extinction. The corresponding distributions of extinction probabilities are very broad since they follow roughly a geometric distribution. It should also be kept in mind that the extinction time

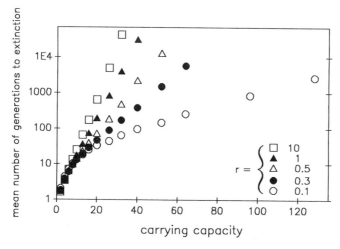

Fig. 3:   Extinction risk due to stochastic sex-ratio and stochastic fecundity under logistic growth as in Figure 2 for various intrinsic growth rates.

depends on the initial population size N_0 which for all results presented here was assumed to be equal to K. Under the logistic population regulation used in this study, the mean extinction time increases monotonically as the initial population size increases and (if $r > 0.5$) rapidly reaches an upper limit, which is almost identical with the values for populations starting at carrying capacity as presented here. Only for very small r and very small initial population size ($N_0 < 4$) is the corresponding extinction time considerably smaller than for $N_0 = K$ (see GABRIEL & BÜRGER submitted).

Mutational Melt-down in Sexual Populations

One might expect that the process of mutational melt-down, as discussed in the previous section for asexual organisms, is not relevant under sexual reproduction because selection and recombination can eliminate bad mutations. But for stochastic logistic population regulation, it is critical to check first how large a carrying capacity has to be in order to produce a sufficiently large effective population size so that selection and recombination will be efficient enough to prevent the accumulation of deleterious mutations. In sexual populations the melt-down process can be effective if the population is temporarily reduced to a size at which the probability of extinction due to sex-ratio imbalance becomes important.

The risk of extinction for sexual populations due to mutational load is demonstrated in Figure 4. Except for very small carrying capacities, the mean times to extinction are reduced by several orders of magnitude relative to the extinction times without mutational load. Preliminary results indicate that this is true also for mutation loads at the lower end of the estimates given in equation (3).

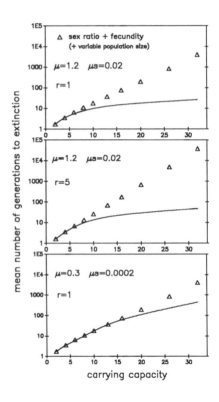

Fig. 4: Survival of sexual populations under mutational load and demographic stochasticity (=solid lines). Intrinsic growth rate r, mutational load (µs), and mutation rates µ are given in each panel. The triangles indicate the corresponding extinction risks without mutational load. (The results for µs = 0.0002 are preliminary.)

Conclusions

The joint action of deleterious mutation and demographic stochasticity has a deleterious synergistic effect which can lead to a mutational melt-down and ultimately to extinction. This process cannot be analyzed in the usual tradition of population genetics which keeps the (effective) population size constant. Evaluation of the effect of deleterious mutation on population viability must be treated in the context of density-dependent population regulation and should allow for demographic variation in family size and sex-ratio.

Since we have neglected risks such as lethal mutation and environmental sources of mortality, times to extinction calculated in this paper are definitely underestimates. For real populations, growth rates and carrying capacities are not constant but more or less dependent on variable abiotic and biotic factors of the environment. Such additional risks are also expected to interact synergistically with unfavorable genetic processes.

Our quantitative results on the impact of mutational load on population extinction are based on estimates of the mutation load derived from experiments which were not primarily designed to measure the mutational melt-down. However, unless the existing estimates of mutational load are greatly exaggerated, the conclusion that the accumulation of deleterious mutations is an important determinant of population extinction seems inescapable for populations with upper size limits of several dozen or smaller. Theoretical studies which combine population genetics with population dynamics are fundamental to the field of conservation biology.

Zusammenfassung

Um die Aussterbezeiten kleiner Populationen abzuschätzen, ist es im Gegensatz zu einer weitverbreiteten Meinung notwendig, auch genetische Faktoren zu berücksichtigen. Dies wird am Beispiel von Mutationen mit nur geringer schädlicher Wirkung demonstriert. Mit Hilfe vorhandener Abschätzungen über Mutations- raten und Mutationseffekte wird untersucht, wie sich solche Mutationslast in Populationen mit dichteabhängigem Wachstum und unter demographischer Stochastizität auswirkt. Die Anhäufung von Mutationen setzt die Überlebenswahrscheinlichkeit der Individuen herab und kann so zu einer zeitweisen Verringerung der Populationsgröße führen, die ihrerseits die Fixierung weiterer schädlicher Mutationen durch genetische Zufallsdrift erleichtert. Diese synergistiche Interaktion wird "mutational melt-down" genannt.

In parthenogenetischen Populationen liegt die "proximate" Ursache für das Aussterben darin, daß auf Grund von Zufallsprozessen keine Nachkommen überleben beziehungsweise geboren werden. Die Wahrscheinlichkeit dafür steigt mit zunehmender Mutationslast und mit abnehmender aktueller Populationsgröße. Für sexuelle Populationen werden als

Referenzpunkte zunächst die Aussterbezeiten unter logistischer Populationsregulation ohne Mutationslast in Abhängigkeit von Wachstumsrate und Kapazität ("carrying capacity") bestimmt. Dabei wirken als stochastische Größen nur Zahl und Geschlecht der Nachkommen. In kleinen sexuellen Populationen können Rekombination und Selektion einen "mutational melt-down" nicht verhindern: unter der Wirkung schwach schädlicher Mutationen verkürzt sich die Aussterbezeit um mehrere Größenordnungen. Entscheidend ist dabei nicht das Ausbleiben von überlebenden Nachkommen sondern die durch temporäres Absinken der Populationsgröße erhöhte Wahrscheinlichkeit, daß nur männliche oder nur weibliche Nachkommen erzeugt werden.

Zur genaueren Abschätzung der Aussterbewahrscheinlichkeit kleiner Populationen sind weitere experimentelle Untersuchungen zur Mutationslast und theoretische Studien zur Interaktion von Populationsgenetik und Populationsdynamik dringend erforderlich.

Acknowledgment

We thank Volker Loeschcke for critical remarks on an earlier version of this manuscript. This work has been supported by a grant from Deutsche Forschungsgemeinschaft to WG. RB acknowledges support by the Austrian Fonds zur Förderung der wissenschaft- lichen Forschung, Projekt P6866. ML was supported by NSF grant 89-11038 and PHS grant RO1 GM36827-O1A1.

References

BELL, G. (1982): The masterpiece of nature: the evolution and genetics of sexuality. Univ. Calif. Press, Berkeley. Bell, G. (1988) Recombination and the immortality of the germ line. J.evol.Biol.1:67-82

CROW, J.F. & M.J. SIMMONS (1983): The mutation load in Drosophila, pp. 2-35. In M. ASHBURNER, H.L. CARSON, & J.N. THOMPSON, Jr. (eds.) The genetics and Biology of Drosophila. Vol. 3C. Academic Press, New York.

EWENS W.J., P.J.BROCKWELL, J.M.GANI, & S.I.RESNICK (1987): Minimum viable population size in the presence of catastrophes. pp 59-68 in SOULÉ (1987).

FELDMAN, M.W. & J. ROUGHGARDEN (1975): A population's stationary distribution and chance of extinction in a stochastic environment with remarks on the theory of species packing. Theor.Pop.Biol. 7:197-207

GABRIEL, W. & R. BÜRGER (submitted) Survival of small populations under demographic stochasticity.

GABRIEL, W. & G.P. WAGNER (1988): Parthenogenetic populations can remain stable in spite of high mutation rate and random drift. Naturwissenschaften 75:204-205

GOODMAN, D. (1987): The demography of chance extinction. pp.11-34 in SOULÉ (1987).

HOPPENSTEADT, F.C. (1982): Mathematical methods of population biology. Cambridge University Press, Cambridge.

HANSON, F.B. & H.C. TUCKWELL (1978): Persistence times of populations with large random fluctuations. Theor.Pop.Biol. 14:46-61

LANDE, R. (1988): Genetics and demography in biological conservation. Science 241:1455-1460

LYNCH, M. (1984): Destabilizing hybridization, general-purpose genotypes and geographic parthenogenesis. Quart.Rev.Biol. 59:257-290

LYNCH, M. & W. GABRIEL (1983): Phenotypic evolution and parthenogenesis. Am. Nat. 122: 745-764

LYNCH, M. & W. GABRIEL (1990): Mutation load and the survival of small populations. Evolution 44: 1725-1737

MAY, R.M. (ed) (1981): Theoretical ecology. 2nd edition, Blackwell scientific publications, Oxford.

MAYNARD SMITH, J. (1978): The evolution of sex. Cambridge University press.

MODE, C.J. (1985): Stochastic processes in demography and their computer implementation. Springer, Heidelberg.

MUKAI, T. (1979): Polygenic mutations, pp. 177-196, in J. N. THOMPSON, Jr. & J.M. THODAY (eds.), Quantitative genetic variation. Academic Press, New York.

MULLER, H.J. (1964): The relation of recombination to mutational advance. Mutat. Res. 1:2-9

NISBET, R.M. & W.S.C. GURNEY (1982): Modelling fluctuating populations. J.Wiley & Sons, Chichester, New York.

RICHTER-DYN, N. & N.S. GOEL (1972): On the extincition of a colonizing species. Theor.Pop.Biol. 3:406-433.

SOULÉ, M.E. (ed) (1987): Viable populations for conservation. Cambridge University Press, Cambridge.

WAGNER, G.P. & W. GABRIEL (1990): Quantitative variation in finite parthenogenetic populations: What stops MULLER'S ratchet in the absence of recombination? Evolution 44: 715-731

WRIGHT, S.J. & S.P. HUBBELL (1983): Stochastic extinction and reserve size: a focal species approach. Oikos 41:466-476

Species Conservation: A Population-Biological Approach
A. SEITZ & V. LOESCHCKE (eds.) © 1991 Birkhäuser Verlag, Basel

The Stabilizing Potential of Spatial Heterogeneity - Analysis of an Experimental Predator-Prey System

H-J. Rennau, Roermonder Straße 137, D-5100 Aachen, Germany

Abstract

1.   The general question is raised: to what extent may a population system be stabilized by spatially fragmenting the given amount of limited resources?

2.   In order to facilitate a systematic approach to this question, two quantitative concepts are introduced: (a) "the degree of spatial heterogeneity" (DSH), (b) the "stabilizing potential of spatial heterogeneity" (SPSH).

3.   In laboratory experiments it was studied how the DSH affected the dynamics of an acarine predator prey system (*Tetranychus urticae* KOCH, *Phytoseiulus persimilis* ATHIAS-HENRIOT). By introducing an appropriate DSH the persistence of the system could be enhanced decisively. This finding was made plausible by further observations of the detailed system behaviour, namely the mean number of local subpopulations as a function of the DSH.

4.   The experiments were simulated by a computer model. The model was successfully validated by the experimental results. Subsequently it was used to analyse the influences which several measurable properties of the individual animals exert on the SPSH.

The simulation studies revealed that the SPSH is largely determined by properties which directly affect local population growth. The developmental rate of the prey and the reproduction rate of the predator (especially if coupled to its feeding rate) proved to be of great importance. On the other hand, properties which determine the dispersal capacity affected the SPSH only to a minor degree.

Introduction

Spatial heterogeneity may influence the stability of populations drastically. The famous experiments by HUFFAKER (1958) and HUFFAKER *et al.* (1963) demonstrated this. In these experiments the dynamics of an acarine predator-prey system were studied. The basic idea was to keep the total amount of supplied resources constant, yet to vary their spatial distribution. The predator-prey system proved to be utterly unstable as long as the distribution was strongly clumped. However, by fragmenting the resources into many small, widely dispersed units, considerable stability could be realized.

This phenomenon may also be relevant with respect to problems of species conservation: habitat areas which offer favourable conditions to a certain species or species community may also be regarded as "supplied resources". A crucial question emerges: To what degree may the stability of a population system be increased by optimizing the degree of resource fragmentation?

Conceptional tools

In order to facilitate a systematic approach, two quantitative concepts will be introduced, the "Degree of Spatial Heterogeneity" (DSH), and the "Stabilizing Potential of Spatial Heterogeneity" (SPSH). These concepts are based on the fundamental assumption that the habitat may be regarded as consisting of discrete patches. Note that this structure is not objectively given but recognized by an act of abstraction. The definition of what constitutes a single patch is therefore arbitrary - it may be a wheat field, a single plant, a pond, etc. The conceptual subdivision of the habitat into patches engenders a subdivision of the population into local populations, each of which is confined to a single patch. The persistence of the population as a whole is compatible with frequent extinctions of local populations, provided the rate of local extinctions is balanced by the rate of local recolonizations.

DSH is qualitatively viewed as a measure for the resistance hampering the transit of animals between different patches. Features of real systems which might be regarded as such a measure are: the distance between adjacent ponds, the reciprocal of an index of leaf contact between neighbouring plants, the width of hedges connecting wood lots, etc. As a fairly general definition, the expected residence time is proposed which migrating individuals spend within a patch.

SPSH is qualitatively viewed as a measure for the degree to which the persistence of a population system may be increased by increasing the DSH, starting at a very low DSH which corresponds to a quasi-homogeneous habitat. (Persistence = time span for which all considered species coexist.) As a quantitative definition I propose the maximally possible increase of the persistence, divided by the persistence under quasi-homogeneous conditions:

$$SPSH = (T^*_{het} - T_{hom}) / T_{hom}$$

where T_{hom} is the expected persistence under quasi-homogeneous conditions, and T^*_{het} the expected persistence that can be maximally achieved by varying the DSH.

Leaning on these concepts and definitions, the introductory question can be reformulated: In which way is the SPSH determined by properties of the system?

This question cannot be answered in relation to population systems in general, it always requires the context of a certain type of population system. In this paper the type "locally unstable predator-prey systems" (LUPPS) is studied more closely. "Locally unstable" means that rapid prey extinction is certain as long as the habitat is small and quasi-homogeneous. Typical examples of LUPPS are many systems of spider mites and predatory mites.

Laboratory Experiments

Materials and Methods

In order to study the effect of the DSH on the dynamics of LUPPS, laboratory experiments were performed. Prey was the spider mite *Tetranychus urticae* KOCH, predator was the mite *Phytoseiulus persimilis* ATHIAS-HENRIOT. The habitat was a closed, cyclic system of petri dishes - the patches - connected by little tubes. Each dish contained a bean leaf on which the spider mites could feed.

The predatory mites were able to migrate from patch to patch, but not the spider mites. The laboratory systems thus differed from natural systems in which both, predator and prey, can migrate from patch to patch.

The experiment was started by placing one leaf in an isolated dish and transferring five spider mite females from stock cultures to the leaf. After three days one female predatory mite was added; at the same time a second dish was attached to the first one and provided with a leaf and five spider mites. Three times per week a further dish was attached to the one previously attached, and each new dish was provided with a leaf and five spider mites from stock cultures. After 14 days the last dish was connected with the first one, rendering the system cyclic. From then on no further dishes were attached. Instead, three times per week the oldest leaf was replaced by a fresh one on which five spider mites were transferred from stock cultures. The periodic addition of spider mites was meant to simulate a periodic immigration into the system.

In similar experiments, besides these so-called "Population Dishes" (PD) also "Empty Dishes" (ED) were employed, to which no prey animals were supplied. The time pattern of introducing PDs to the system and of replacing old leaves by fresh ones was the same as described. The EDs were introduced along with the PDs.

EDs consituted a "resistance" which impeded the predator's migration from PD to PD. The DSH was varied by varying the number of EDs: Four levels of DSH were used by combining 6 PDs with 0;3;6;9 EDs. The respective sequences of dishes were: PPPPPP; PPEPPPEPPE; PEPEPEPEPE; PEPEEPEPEEPEPEE. All setups were cyclic, i.e. the last dish was connected to the first one. The experiments were replicated four times (twice, using 9 EDs).

Results and Interpretation

The observed influence of spatial heterogeneity on the population dynamics is summarized in Table 1. Under conditions of very low heterogeneity (no EDs employed, corresponding to DSH = 0.43d) the population dynamics resembled those in a homogeneous system: only one single predator-prey oscillation occurred, and rapid prey extinction was almost certain. Increasing the DSH slightly (3 EDs employed) made longer coexistence of prey and predator possible (occurring in 2 replicates). However, again in all replicates prey extinction was the final outcome. A further increase of the DSH (6 EDs employed) allowed for long-term coexistence of prey and predator: in 3 out of 4 replicates both species were still present after 100 days. The highest DSH tested (9 EDs employed) produced a less persistent system, and now it was predator extinction which ended the coexistence.

These results can be summarized as follows: (a) the probability of prey extinction is a monotonously decreasing function of the DSH; (b) a considerable persistence of the predator-prey system can be induced by imposing an intermediate DSH; when the DSH is low, the persistence is severely restricted by impending prey extinction; at too high a DSH the persistence is limited by the tendency for predator extinction.

Further effects of spatial heterogeneity were observed. (a) Increasing the DSH also led to an increased mean persistence of local populations; this effect was strong (weak) with respect to the prey (predator). (b) The mean number of local prey populations was a monotonously increasing function of the DSH. (c) The mean number of local predator populations reached a maximum at an intermediate DSH (6 EDs).

Tab. 1:   The influence of spatial heterogeneity on the dynamics of an experimental predator-prey system. The habitat is a cyclic system of petri dishes, interconnected by little tubes. DSH1 (2;3;4) denote four increasing Degrees of Spatial Heterogeneity created by interspersing 0 (3;6;9) "Empty Dishes" between 6 "Population Dishes". Four replicates, 1-4, two at DSH4.

	T: extinction of prey P: extinction of predator C: coexistence				persistence (d)				
	1	2	3	4	single values				mean
					1	2	3	4	
DSH1	T	T	T	T	21	23	26	82	38.0
DSH2	T	T	T	T	26	26	77	91	55.0
DSH3	P	C	C	C	56	>100	>100	>100	>89.0
DSH4	P	P	-	-	20	71	-	-	45.5

The experimental results may be interpreted as follows. In a quasi-homogeneous system there is no stability because the prey is too "weak" - its rapid extinction is certain. Increasing the DSH results in a "strengthening" of the prey and, sooner or later, a "weakening" of the predator (both effects are reflected by changes in the mean number of local populations). Spatial heterogeneity permits a substantial stabilization of a LUPPS if it can be applied in such a dose which effects a considerable "strengthening" of the prey but not a dramatic "weakening" of the predator. In the experimental systems this "balancing act" was successfully performed by introducing 6 EDs.

Can the observed high SPSH be generalized with respect to LUPPS? Here it must be remembered that the experimental systems had highly artificial features (no prey migration; prey supply from without; resource supply from without). The question is whether the observed system behaviour may have been conditioned by these unnatural properties. Another interesting question is: how sensitively does the SPSH respond to changes in the life cycle properties and the dispersal capacity of the animals?

Simulation experiments

Method

A simulation model was developed to provide a bridge between the experimental systems and nature. It was intended to reveal how the experimental systems would behave if the unnatural features were avoided. First, of course, it had to be shown that the model succeeded in simulating the actual experiments with reasonable realism. The model validation (described in RENNAU 1989) produced in fact quite satisfactory (though not perfect) results.

After validation, the model was used to study the relationship between spatial heterogeneity and stability in a systematic way. All parameters of the species (reproduction rates, developmental rates, etc.) were retained. However, in the simulated system the prey was allowed to migrate from patch to patch. It was arbitrarily assumed that the dispersal capacity of the prey was identical to that of the predator. The simulated system comprised 40 patches arranged in a 5x8 pattern. During the simulation experiments there was no interference from without once the experiment was started. For a detailed description of the model and the simulated system see RENNAU 1989 (p. 37-82, 94-97).

Results: DSH varied, properties of the animals unchanged

In a first series of simulations, it was studied how the DSH influenced the system behaviour when the properties of the animals remained unchanged. (The DSH was varied by varying the "Search Rate" of the animals, defined as the number of patches searched by unit time, i.e. 1/DSH.)

It was not possible to achieve a substantial stabilization of the predator-prey system simply by varying the DSH (Fig. 1a). The maximum of the "persistence-DSH-curve" was below 100 days. So the SPSH of the simulated system turned out to be rather small, contrary to the experimental system (in which 3 of 4 replicates were stopped after 100 days of coexistence). The effect of the DSH on the probability of final prey extinction was the same as in the experimental system: an increase of the DSH decreased the probability of prey extinction.

The population dynamics process in a heterogeneous (patchy) system may be divided into two subprocesses: (a) INTRApatch dynamics, i.e. the growth of local populations; (b) INTERpatch dynamics, i.e. the exchange of migrants between the patches. To characterize the INTRApatch process I used the "Net Production of Local Populations" (NPLP): the total number of emigrants leaving a local population, minus the total number entering it (both numbers averaged over all local populations). In order to characterize the INTERpatch process, I used the "Probability of Immigration" (P(I)): the fraction of all emigration events that lead to subsequent immigration events. P(I) thus measures an emigrant's probability to survive migration.

Fig. 1b shows how the INTRApatch processes of prey and predator responded to variations of the DSH. An increase of the DSH had a marked positive effect on the local prey production but hardly any influence on the local predator production. Such a difference between prey and predator did not occur concerning the INTERpatch process: the negative effect of the DSH on the Probability of Immigration was quite similar for both populations (Fig. 1c).

Concerning the prey, the moderate negative effect of the DSH on the INTERpatch process was more than compensated by the positive effect on the INTRApatch process. Therefore, an increase of the DSH resulted in an increased mean number of local prey populations (Fig. 1d). Concerning the predator, there was no such positive effect of the DSH on the INTRApatch process, and therefore the negative effect on the INTERpatch process caused an overall decline in the number of local population (Fig. 1d).

The relatively small SPSH may be interpreted as follows. While it was possible to "strengthen" the prey by increasing the DSH, this was coupled to a "weakening" of the predator to such an extent that a substantial stabilization of the two-species-system was excluded. A better stabilization, it appears, would require changes in the properties of the animals themselves.

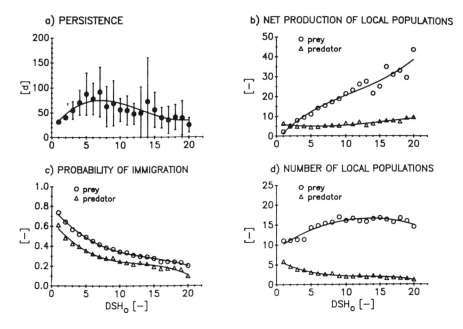

Fig. 1:   The influence of spatial heterogeneity on the dynamics of a simulated predator-prey
            system. The habitat comprised 40 patches, arranged in a 5 x 8 pattern. For each tested
            Degree of Spatial Heterogeneity (DSH), 10 simulation replicates were run.
            DSH_0 : normalized Degree of Spatial Heterogeneity ($DSH_0 = DSH / 0.43d$).

Results: INTRApatch properties of the animals varied

The properties of the animals may be divided into INTRApatch properties - directly
affecting the INTRApatch process - and INTERpatch properties, directly affecting the
migration process.

The following INTRApatch properties were considered:

- reproduction rate of prey and predator
- developmental rate of prey and predator
- emigration rate of prey and predator
- feeding rate of the predator
- reproduction and feeding rates of the predator, coupled

The parameters were varied one by one. For each parameter 4-8 values were tested, and 10 simulation replicates run per value. All parameter variations were repeated at 4 different DSHs (0.43d x (1; 2; 4; 8)).

Table 2 summarizes the results. The stability of the system was markedly enhanced, (a) by increasing the developmental rate of the prey, and (b) by decreasing the reproduction rate of the predator, especially if this decrease was coupled to a decrease of its feeding rate.

As the parameter variations were repeated at 4 different DSHs, it was possible to derive rough estimates of the SPSH as a function of the parameter values. Like the persistence at a given DSH, also the SPSH responded sharply to the developmental rate of the prey and to the coupled reproduction and feeding rates of the predator. The other INTRApatch properties proved to be surprisingly irrelevant with respect to the SPSH.

Tab. 2:   The influence of INTRApatch properties of the animals on the stability of a simulated predator-prey system, observed at 4 different Degrees of Spatial Heterogeneity.
The left part of the table gives the Stabilizing Potential of the Parameter (SPP), defined by SPP = $(T^* - T_0) / T_0$ where T_0 is the expected persistence resulting when the reference value of the parameter is chosen, and T^* the expected persistence that can be maximally achieved by varying the parameter. The reference values refer to T. urticae (prey) and P. persimilis at 25 °C. DSH_0 = normalized Degree of Spatial Heterogeneity (DSH_0 = DSH/0.43d).
RR = reproduction rate; DR = developmental rate; ER = emigration rate ; FR = feeding rate; RRFR = reproduction & feeding rates, coupled

		STABILIZING POTENTIAL at DSH_0				IMPORTANCE	stabilization by increase (+) or decrease (-) of par. value? DSH_0			
		1	2	4	8		1	2	4	8
PREY	RR	0.1	0.1	0.1	0.7	*	-	-	-	+
	DR	>0.0	>0.6	>2.3	>0.4	****	+	+	+	+
	ER	0.1	0.0	0.5	0.7	*	+		+	+
PRED.	RR	3.3	7.6	2.0	0.9	****	-	-	-	+
	RRFR	2.8	2.6	2.7	>8.9	****	-	-	-	-
	FR	0.3	0.2	0.5	0.4	*	+	+	+	-
	DR	1.4	0.6	0.9	0.8	**	-	-	-	-
	ER	0.0	0.0	0.0	0.8	*				+

Results: INTERpatch properties of the animals varied

As INTERpatch properties, 4 parameters per species were studied:

- the "search rate": the number of patches searched per unit time;

- the "search perseverance": the time span survived without food;

- the "search lag": the mean time span between the depletion of local food resources and emigration;

- the "probability of aerial dispersal": the probability that a patch to patch movement leads not to one of the adjacent patches but to some randomly chosen patch of the system .

The dispersal capacity of the prey was of little importance for the stability of the predator-prey-system. An exception is the search perseverance: increases have a markedly stabilizing effect, but this only under the condition of an intermediate DSH.

The dispersal capacity of the predator affected the stability more than that of the prey. At a DSH greater (smaller) than the optimal DSH, some stabilization was achieved by an appropiate increase (decrease) of the dispersal capacity. However, it must be stressed that variations of the predator's dispersal capacity permitted roughly the same degree of stabilization as could have been achieved by varying the DSH. Consequently, the SPSH was practically independent from the dispersal capacity of the predator. Nevertheless, this capacity largely determined which DSH did in fact permit maximal stability: an increased (decreased) dispersal capacity caused an increased (decreased) "optimal" DSH.

Integration of the results

Our original question was: To what degree may the stability of a system be enhanced by optimizing the degree of resource fragmentation? Fig. 2 attempts to integrate the results reported above.

The persistence of the LUPPS in a quasi-homogeneous habitat (Fig. 2: distance A) is, evidently, determined by the INTRApatch properties of prey, predator and resources, namely the rates of reproduction, development, abiotic mortality and feeding and the rate of resource regeneration.

The maximal persistence (distance A+B) that may be achieved by varying the DSH does likewise depend on these INTRApatch properties (compare Table 2), and also on the search perseverance of the prey, but hardly on other aspects of the prey dispersal capacity or on any aspect of the predator dispersal capacity. Among those INTRApatch properties of prey and predator, the developmental rate of the prey and the reproduction rate of the predator exert a superior influence.

As stated, the predator dispersal capacity is rather irrelevant with respect to the distance A+B; it does, however, largely determine the distance C, i.e. the DSH permitting maximal persistence: the better the predator dispersal capacity is, the higher that "optimal" DSH was observed to be.

In conclusion, with respect to LUPPS, these two rules of thumb are proposed:

(a)   A high SPSH is favoured by a high rate of prey development and by a low rate of predator reproduction, further by a high search perseverance of the prey.

(b)   The DSH permitting maximal persistence is positively related to the dispersal capacity of the predator.

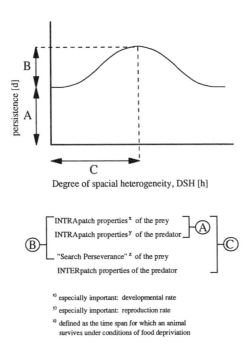

Fig. 2:   Attempt at an integration of the simulation results: the stability of a "locally unstable predator-prey system" as determined by spatial heterogeneity and properties of the animals.

Discussion

The reported results were obtained for one specific system characterized by certain species exposed to certain abiotic conditions, by a certain number of patches, etc. One may legitimately ask whether the results can be considered representative of the system type "Locally Unstable Predator-Prey System".

Without comparable studies on other specific systems any generalizations are of course very hypothetical. *Understanding* the underlying causal processes should help to provide a reasonable basis for generalizations as well as for grasping in which way certain aspects of the system behaviour are conditioned by particular features.

How to approach such a causal understanding? To my mind it is helpful to regard stability as *"global system behaviour"* to be distinguished from a *"detailed system behaviour"* (comprising e.g. the Net Production of Local Populations, the Probability of Immigration, the mean Number of Local Populations) and to relate the phenomena observed on both levels. Thus, comparing two actual systems, differences of the global behaviour (e.g. of the SPSH) may be recognized as resulting from striking differences of the detailed behaviour. And the latter ones may in turn be interpreted as caused by certain differences of basic properties of the system (e.g. life cycle properties). Observing the detailed behaviour can thus provide data that serve as stepping stones, helping to bridge the space between the basic properties of the system (namely the measurable properties of the individual animals) and the ultimate effect, i.e. stability. Maybe it is a key problem to define the detailed system behaviour: to arrive at an appropriate choice of which quantities are to be included.

Zusammenfassung

1.   Es wird die allgemeine Frage gestellt: in welchem Maße kann ein Populationssystem dadurch stabilisiert werden, daß eine begrenzte Ressource räumlich fragmentiert wird?

2.   Um eine systematische Bearbeitung der Frage zu erleichtern, werden zwei quantitative Konzepte eingeführt: (a) der "Grad räumlicher Heterogenität" (DSH), (b) das "Stabilisierungspotential räumlicher Heterogenität" (SPSH).

3.   In Laborexperimenten wurde beobachtet, wie der DSH die Dynamik eines Räuber-Beute-Systems von Spinn- und Raubmilben (*Tetranychus urticae* KOCH, *Phytoseiulus persimilis* ATHIAS-HENRIOT) beeinflußt. Durch Einführung eines geeigneten DSH konnte die Stabilität des Systems drastisch erhöht werden. Dieses Ergebnis wurde plausibel durch weitere Beobachtungen zum Systemverhalten, namentlich der mittleren Anzahl lokaler Teilpopulationen als Funktion des DSH.

4.   Die Experimente wurden durch ein Computermodell simuliert. Das Modell konnte durch die experimentellen Ergebnisse erfolgreich validiert werden. Danach wurde es als ein Werkzeug benutzt, um den Einfluß zu untersuchen, den verschiedene meßbare Eigenschaften der Tiere auf das SPSH ausüben.

Es zeigte sich, daß das SPSH vor allem durch jene Tiereigenschaften bestimmt wird, welche unmittelbar das lokale Populationswachstum beeinflusssen. Namentlich die Entwicklungsrate der Beute und die Reproduktionsrate des Räubers (diese besonders, wenn an seine Freßrate gekoppelt) erwiesen sich als sehr wichtig. Dagegen beeinflußten Tiereigenschaften, die das Dispersionsvermögen bestimmen, das SPSH nur wenig.

References

HUFFAKER, C.B. (1958): Experimental studies on predation: dispersion factors and predator-prey oscillations. Hilgardia 27: 343-383.

HUFFAKER, C.B.; SHEA, K.P.; HERMAN, S.G. (1963): Experimental studies on predation: complex dispersion and levels of food in an acarine predator-prey interaction. Hilgardia 34: 305-320.

RENNAU, H. (1989): Der Einfluß räumlicher Heterogenität auf die Stabilität eines lokal instabilen Räuber-Beute-Systems von Spinn- und Raubmilben: experimentelle Analyse und Simulation. PhD-Thesis, RWTH Aachen.

Species Conservation: A Population-Biological Approach
A. SEITZ & V. LOESCHCKE (eds.) © 1991 Birkhäuser Verlag, Basel

Ecological Risk Analysis for Single and Multiple Populations

H. R. Akçakaya [1] and L. R. Ginzburg [2]

[1] Applied Biomathematics, 100 North Country Road, Setauket, NY 11733, USA
[2] Department of Ecology and Evolution, State University of New York at Stony Brook, Stony Brook, NY 11974, USA

Abstract

Three mathematical models for assessing extinction risks are introduced. The first is a Monte Carlo model that describes the growth of an age-structured population under environmental and demographic stochasticity. The parameters are the mean value, variation, distribution and correlations of life history traits such as survival, fecundity and migration, and density dependence in recruitment. Probabilities of population decline and extinction are computed as a way to evaluate ecological risk. The other two models describe the dynamics of multiple population systems (metapopulations) given their spatial structure and the rate of migration among populations. One of these metapopulation models is based on the occupancy of habitat patches, and the other on the population dynamics within local populations. These models evaluate the risk of extinction of the whole metapopulation as well as the local populations.

Three computer programs (RAMAS library) for building single and multiple population models are described. The programs are used to build age-, stage- and spatially-structured models. They provide summaries regarding the expected number of individuals as a function of time, and the probability that the population size will fall below a threshold.

Introduction

An unfortunate corollary of human existence on the earth is the large decrease in biological diversity caused by our activities. The current rates of extinction is several orders of magnitude greater than the background rate of about one species per year over geological time. Conservation biology is a new field concerned with the maintenance of biotic diversity. How large should natural reserves be? Is it better to have one large reserve or several smaller reserves? How resilient will a given population be to natural variability in its environment, and how will it respond to changes brought by humans? One way of addressing such questions is through the language of risk assessment, which seeks to express extinction risks in probabilistic terms.

Our approach combines population ecology with the methods of risk assessment. The idea is to emulate the population's natural variability in a model of population growth using observed means and variances of biological parameters such as survival and fecundity. Population trajectories showing the changes in abundance through time can then be simulated numerically using computers. The "natural" variation included in the model causes each population trajectory to be unique even when given the same initial conditions, and the range of outcomes observed among the population trajectories can be used to estimate the extinction probabilities. Estimates of extinction risks, and the management decisions which take such risks into account, may then be based on observations of the distributions of demographic parameters.

In this paper we introduce single and multiple population models that use the risk assessment approach. In the next section we briefly describe age- and stage-structured models to simulate the dynamics of a single population. Then we discuss additional factors that affect extinction risks in multiple population systems and introduce a Markov model that incorporates these factors and is based on the occupancy status of habitat patches. We demonstrate the use of this model with a specific example. Finally, we discuss the limitations of occupancy models and introduce a different approach based on the population dynamics at each patch.

Ecological risk assessment for single populations

The size and structure of biological populations change continually and sometimes radically, even without any human influence. Until recently, environmental scientists discussed an environmental impact as though it were a single fixed quantity that need only be measured and controlled. This view has lately become more sophisticated to recognize the essential nature of random variation in ecological phenomena. Now, we are interested in computing the risk of unwanted consequences such as a species going extinct or falling to some low level of abundance. Such ecological risks are measured as probabilities and we can express an estimated risk in probabilistic statements like "There is a 5% chance that the size of the population will drop by 20% or more within the next few years". In fact, there is always some non-zero probability that the population's abundance will drop by 20% even if the influence of humans could be completely erased. The probability of this occurring can be considered the natural background level of risk. Problems in conservation biology and environmental management require an understanding of the natural risks and involve measures to control increases in the risks due to human influences.

One approach to assessment of extinction risks is to utilize species-specific information in models of population dynamics. Such models may incorporate factors such as density dependence and demographic and environmental stochasticity (LEWONTIN & COHEN 1969, LEVINS 1969, MAY 1973, ROUGHGARDEN 1975, BOYCE 1977, TULJAPURKAR & ORZACK 1980, GINZBURG et al. 1982, LANDE & ORZACK 1988). When other factors such as age-

Tab. 1:   Parameters of RAMAS that can be specified by the user

Scalars	Age-structured vectors	Functions
Sex ratio	Fecundity	Environmental/Demographic stochasticity
Fecundity variation	Natural survival	Density dependence
Adult survival variation	Migration	Fecundity distribution
Juvenile survival variation	Initial abundance	Juvenile survival distribution
Migration variation		Adult survival distribution
Time to run		Migration distribution
Number of replications		Correlations among fecundity, survival and migration

structure and correlated variation of survival and fecundity between age classes are also incorporated into stochastic population models, analytical solutions are not possible and numerical simulations are used (SHAFFER 1983, SHAFFER & SAMSON 1985, GINZBURG *et al.* 1984). A computer program, RAMAS/age, was designed with this approach to assess the risks of population extinction (FERSON and AKÇAKAYA 1990). RAMAS/age is a numerical simulator of age-structured population dynamics for the IBM PC family of personal computers. It allows users to build age-specific population models by specifying the input parameters listed in Table 1 (FERSON *et al.* 1989).

RAMAS/age answers questions regarding the expected number of individuals in a specified age class after a certain number of years along with the reliability of this estimate and the probability that the population size will fall below a specified threshold during or at the end of a specified time period, which is the quasiextinction risk (GINZBURG *et al.* 1982). Since the choice of the quasiextinction threshold is subjective (e.g., it can be picked as a critical population size that is important in a biological or an economic sense), RAMAS computes the quasiextinction risk as a function of threshold. Applications of RAMAS/age include an analysis of the effects of density dependence on extinction risks (GINZBURG *et al.* 1990), and an analysis of the response of Hudson River striped bass populations to fishing and power plant mortality and recovery of bluegill sunfish populations after a population crash due to pollution (FERSON *et al.* 1990).

Some life-histories such as those of plants or insects, or species with sexual dimorphisms and behavioral castes cannot be adequately described by age-structured models. For these species, stage (rather than age) of the organisms determine their demographic characteristics. A computer program, RAMAS/stage, has recently been developed to incorporate information on stage-structure and on the effect of environmental factors (FERSON 1990).

Determining the Risk of Species Extinction

The methodology described in the previous section has been developed for assessing the risk of extinction for a single population. But most species do not exist as single populations in nature. Usually there are several populations (i.e., a metapopulation) that are either isolated from each other or that exchange a limited number of individuals. Understanding the link between population extinction and species extinction has obvious practical implications for predicting the fate of species under anthropogenic impact. Factors that affect population extinction are the population's size and age structure, its life history parameters, demographic and environmental stochasticity that cause variation in these parameters, and the correlations among life history parameters within populations. Species extinction risk depends on all these, plus other factors that describe interactions among the populations, including how many populations there are, the correlations of environmental conditions they experience, and the migration that leads to recolonization of locally extinct populations.

Having multiple independent populations decreases the risk for the ensemble, but having low abundances at each site increases the risks of extinction for each of the populations. Geographical proximity usually allows migration and potentially recolonization, but it can also result, because of similarity in environmental patterns experienced by the populations, in their behaving as one population with respect to stochastic environmental changes and thus increase the risk that they will be lost simultaneously. These trade-offs and other complexities in the relation between population and species extinction probabilities, prevent our making unconditional generalizations about whether a particular management strategy will increase or decrease overall extinction risks. One important practical question related to this discussion is the design of nature reserves. The question of whether a single large or several small reserves give more protection to a threatened species has been a major subject of debate.

Three types of models have been used to address these questions. The first is the theory of island biogeography (MACARTHUR and WILSON 1967), which does not have much practical value in the management and conservation of wildlife populations (see SIMBERLOFF & ABELE 1982, MARGULES et al. 1982, BURGMAN et al. 1988 for reviews). The second is a metapopulation model based on the occupancy status of habitat patches, and the third type of model is based on explicit modeling of population dynamics in each patch.

LEVINS (1970) proposed the first occupancy model which described the rate of change in the number of occupied habitat patches. The major assumptions of this model are that all local populations are identical and there is no spatial structure. QUINN and HASTINGS (1987) derived a formula based on the single population model developed by LEIGH (1981) and suggested that the best way to divide the reserves is to have $m = \sqrt{N}$ reserves, where N is the total number of individuals. As GILPIN (1988) pointed out, their formulation did not include the consideration of environmental correlations among localities and migration and recolonization between populations. The question of whether a single large or several small reserves offer more protection for species (the SLOSS argument) can only be answered by taking these two important factors into account. Recently, DOMBROVSKII and TYUTYUNOV (1987) and HARRISON and QUINN (1989) modeled metapopulation dynamics in correlated environments, but these models described correlation with a single parameter, assuming that all pairs of populations have the same correlation of extinction probabilities. Furthermore, they assumed that the probability of recolonization from any population to any other was the same. These models do not incorporate a full spatial structure, since they implicitly assume that all the local populations are at the same distance from each other. The model we describe in the next section is an occupancy model that incorporates a full spatial structure, and unequal population sizes.

A Markov Model of Species Extinction

In this model the species-specific information is summarized in the parameters which are estimated from a more detailed simulation incorporating any level of complexity but is limited to a single locality. The suggested approach is to solve the problem in two steps, first by generating local extinction parameters from a simulation model (such as RAMAS/age) and then analyzing the total problem in a metapopulation model.

Variables and parameters

We used a probabilistic metapopulation model based on a Markov process of transition between states, which represent the occupancy status of habitat patches. Each state is described by a time-dependent variable that represents the probability that a particular combination of populations remain extant to the end of time t. Consequently the number of states is equal to 2^n where n is the number of populations. Here we describe the model for the case of two populations, in which case there are four variables in the model. These four variables that describe the probability of each of the four states are:

$S_{11}(t)$: Both populations are extant at time t

$S_{10}(t)$: Population 1 is extant, and population 2 is extinct at time t

$S_{01}(t)$: Population 2 is extant, and population 1 is extinct at time t

$S_{00}(t)$: Both populations are extinct at time t

At each time period, the transition from each of these states to others are described by transition matrices. These transition matrices are based on the parameters of the model. There are two sets of parameters in this model. The first set consists of *conditional probabilities* of population extinction in a given year (P_{ij}). These are the probabilities that a particular population goes extinct or survives in a single time period, given that the other population goes extinct or survives. These are given in the following table:

Population 1

		Goes extinct	Survives	Total
	Goes extinct	P_{00}	P_{10}	P_{*0}
Population 2	Survives	P_{01}	P_{11}	P_{*1}
	Total	P_{0*}	P_{1*}	1.0

The column and row totals show the marginal probabilities. For example P_{0*} is the probability that population 1 goes extinct in one time period, regardless of whether population 2 goes extinct or survives that time period. The set of four conditional probabilities describe the degree of dependence between the survival of the two populations. There are an infinite number of combinations of these conditional probabilities that will give the same marginal totals, each of which correspond to a particular degree of association or dependence between the two populations. For example, for a marginal probability of extinction of 0.2 for both populations, the following tables represent three such combinations.

A	B	C

0.2	0	0.2
0	0.8	0.8
0.2	0.8	1.0

0.04	0.16	0.2
0.16	0.64	0.8
0.2	0.8	1.0

0	0.2	0.2
0.2	0.6	0.8
0.2	0.8	1.0

Combination A represents complete dependence or association between the two populations. The marginal extinction probability of each population is equal to P_{00}, i.e., they can only go extinct together. The other extreme of negative association is represented by the combination C, where P_{00} is zero, i.e., if one population goes extinct, the other survives. Combination B represents independence of the two populations: in this case the product of diagonal elements (0.04 and 0.64) is equal to the product of off-diagonal elements (0.16 and 0.16), and the probability of both populations going extinct per unit time (0.04) is the product of the marginal probabilities of each population going extinct per unit time (0.2). It represents a situation in which environmentally induced variation in the two populations is uncorrelated. The conditional probabilities for real populations will probably be between A and B, and will not extend to C.

The second set of parameters describes the migration between the two populations that results in the successful recolonization of an extinct population by the migrants from the other population. The parameters M_{ij} represent the probability of migration and recolonization from population i to population j per unit time. M_{ij} need not be equal to M_{ji}, thus the migration between two populations may be asymmetric.

Transitions

The probability of each state changes at every time period as a result of events that effect survival and as a result of recolonization of locally extinct populations. These changes are described by two transition matrices. Each element of the matrices represents the probability of transition from one particular state at time t-1 (columns) to a particular state at time t (rows). The first matrix is based on the conditional probabilities and their marginal totals described above and represents the events that determine the extinction and survival of populations. For example if the species is in state S_{10} at time t-1 (population 1 extant, population 2 extinct), then in the next time step it will either stay at the same state (i.e., population 1 will survive) with probability P_{1*}, or it will go to state S_{00} (i.e., population 1 will go extinct) with probability $1-P_{1*}$ ($=P_{0*}$).

Transition between states: Extinction / Survival

		From state at time t - 1			
		S_{11}	S_{10}	S_{01}	S_{00}
	S_{11}	P_{11}	0	0	0
To state	S_{10}	P_{10}	P_{*1}	0	0
at time t	S_{01}	P_{01}	0	P_{*1}	0
	S_{00}	P_{00}	$1-P_{1*}$	$1-P_{*1}$	1

The second matrix is based on recolonization probabilities, and has the same properties. For example if the species is in state S_{10} at time t-1 (population 1 extant, population 2 extinct), then in the next time step it will either go to state S_{11} (by recolonization from population 1 to population 2) with probability M_{12}, or it will stay in state S_{10} (i.e., no recolonization) with probability $1-M_{12}$.

Transition between states: Recolonization

		From state at time t - 1			
		S_{11}	S_{10}	S_{01}	S_{00}
	S_{11}	1	M_{12}	M_{21}	0
To state	S_{10}	0	$1-M_{12}$	0	0
at time t	S_{01}	0	0	$1-M_{21}$	0
	S_{00}	0	0	0	1

Note that (1) since these are transition matrices their columns add up to one, (2) a species at state S_{00} (all populations extinct) can only remain at that state, (3) migration has no effect if the species is in state S_{11} (all populations extant), (4) extinction events (first matrix) can either decrease or not change the number of extant populations, whereas migration (second matrix) can either increase or not change the number of extant populations, and (5) if all migrations are zero, the second matrix is reduced to an identity matrix (with diagonal elements equal to one and off-diagonal elements equal to zero) and therefore does not change the state.

The probability of each state at time t is calculated by the multiplication of the vector of state variables at the previous time period, $S(t-1)$, with the transition matrices P and M:

$$S(t) = P \ M \ S(t-1)$$

The order of multiplication (in this equation first with M then with P) did not make an important difference in the results, especially after several time periods. The species extinction probability is given by the state variable $S_{00}(t)$, which is the probability that both populations will be extinct at time t. The same analysis can also made with three populations, in which case there are 8 states and the state $S_{000}(t)$ gives the probability that all three populations (hence the species) will be extinct at time t. The transition matrices for three populations are considerably more complex, so we will not give them here even though we will use a three population example in the next section.

Results

We will demonstrate the use of the model and the types of predictions that can be made with a specific example from an endangered species, the Mountain Gorilla (*Gorilla gorilla beringei*). We used data on Mountain Gorilla populations in the Virunga mountains of south-west Uganda, which live in three populations more or less isolated from other gorilla populations. We used information given by HARCOURT *et al.* (1981) and WEBBER and VEDDER (1983) on these populations to calculate age-specific survivorship and fecundity values for 5-year periods. We then estimated the *annual* probability of extinction for a *single* population using RAMAS/age. We studied the effect of environmental correlations and migrations among the three populations by using a series of conditional probabilities representing a range of correlations (r) from zero (independence) to one (complete dependence), and by using a range of migration and recolonization probabilities from 0% to 1% per year.

Figure 1 shows the extinction probability as a function of time under environmental fluctuations with different levels of correlation for the 3-population system. In this figure, migration and recolonization probabilities are zero, and a 1-population system (one large population with 3 times the size of a single population) is also shown for comparison. Similarly, Figure 2 shows the effect of recolonization. In this figure, the environmental fluctuations are uncorrelated (*r*=0), and all six recolonization probabilities among three populations are equal. The reason is lack of information on correlations and migrations. However, the advantage of this model over the previous occupancy models is that if such data were available it could incorporate them.

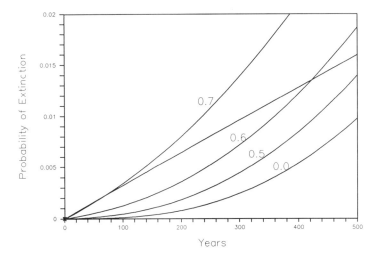

Fig. 1:   Probability of species extinction of the three-population system at different levels of environmental correlation from 0 to 0.7 (recolonization probability = 0).
The straight line represents a single large population

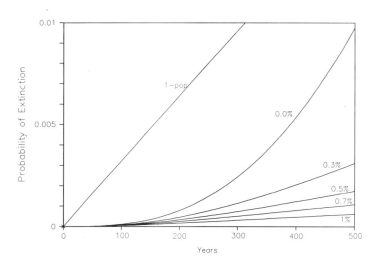

Fig. 2:   Probability of species extinction of the three-population system at different levels of recolonization probability from 0% to 1% (environmental correlation = 0).

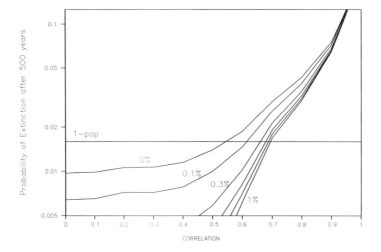

Fig. 3:   Probability of species extinction after 500 years of the three-population system at different levels of recolonization probability (from 0% to 1%) and environmental correlation (from 0 to 1). The straight line represents a single large population

Figure 3 shows the combined effects of recolonization and environmental correlation for the 3-population system. In this figure, the probability of species extinction after 500 years is given as a function of 11 levels of correlation from zero to one and six levels of recolonization probability from 0% to 1%. The horizontal line shows the extinction probability of a 1-population system after 500 years (note that the vertical scale is logarithmic).

Discussion

The results of this model can be interpreted as the sensitivity of extinction probabilities to the unknown factors about this population: correlation of environmental variation among localities and the probability of recolonization among populations. The data on which the analysis is based come from a population with insufficient information about its subdivisions and about the life history parameters in the subpopulations. Therefore the interpretation of results is necessarily a hypothetical exercise, aimed at a qualitative analysis of trade-offs between various factors affecting metapopulation dynamics, and a demonstration of how the available information on multiple population systems can be incorporated into the proposed model, rather than quantitative predictions about the future of a specific population. However, this analysis demonstrates that given sufficient information about the spatial and demographic characteristics of a metapopulation, the proposed model can make quantitative predictions about its risk of extinction under different conditions.

Our results show that a high correlation between three small populations makes them more vulnerable to extinction than the single large population. For example, if there is no migration between the populations, the 3-population system in moderately correlated environments ($r > 0.7$) have no advantage over the 1-population system (Figure 1). With lower correlations, the multiple population systems initially have an advantage which decreases with time; the curves for single and multiple population systems cross, indicating that multiple populations eventually have a higher extinction risk. This is even true in the case of uncorrelated environments, if the recolonization probability is zero or small (Figure 2). In Figure 2, a recolonization probability of less than 0.3% seems to be enough to insure a long-term advantage for the 3-population system over the 1-population system, given the former has zero environmental correlation between its three populations. When the populations are in correlated environments, the threshold recolonization probability (which is the minimum in order for a multiple population system to have a lower risk than the single population system) is higher. Figure 3 shows that for a 3-population system to have lower extinction risk than the single population system, r (correlation) must be less than 0.5 if m (recolonization probability) is zero; r must be less than 0.6 if $m= 0.1\%$, and less than 0.7 if $m= 1\%$. Thus there seems to be a minimum rate of recolonization (the exact value of which depends on the correlation between populations) necessary in order to make a multiple-population system less vulnerable to extinction than a single population of the same total size.

The two-step approach we used is a practical way of taking information on the biology of a species and on the spatial structure of its metapopulation into account. The advantages of this approach over the other occupancy models are: (1) a specific correlation coefficient and a specific migration rate for each pair of populations introduce a full spatial structure that the models discussed above lack; (2) the local populations are not assumed to be identical; the total extinction probability may be different for each population, summarizing the differences in their carrying capacities and population dynamics; (3) it is possible to compute the risk of extinction for any subset of populations, in addition to the risk of extinction of the entire metapopulation.

Despite these advantages, this model shares some of the limitations that are intrinsic to all probabilistic models. The most important limitation of all the metapopulation models discussed so far pertains to parameter estimation. Although parameter estimation is the most crucial and limiting step for any population model, it is especially difficult for probabilistic models. Measuring probabilities of extinction is almost impossible, thus models using extinction probability as a parameter are generally difficult to apply to real cases. The Markov model described here tried to overcome this problem by estimating its parameters from another model which uses life history data as parameters and extinction probabilities as output. In this way it was also possible to simulate the effect of correlation between extinction probabilities. This effect was analyzed as the sensitivity of the total (species) extinction probability to the degree of correlation among local extinction probabilities per unit time. Clearly, it would be much more difficult to guess the actual degree of correlation in a real metapopulation.

An additional limitation of this Markov model is the number of parameters. As the number of patches that can be occupied by local populations increase, the parameter estimation process becomes too complicated for practical purposes. On the other hand, for metapopulations composed of a small number of local populations the model provides a fast and relatively realistic method for comparing extinction probability of different metapopulations. For larger metapopulations a more convenient method is a simulation approach based on local population dynamics. We recently developed such a metapopulation model (and an interactive software RAMAS/space used for building spatially-structured population models; AKÇAKAYA and FERSON 1990). This model simulates the growth of each local population with a stochastic model that can include density dependence (such as overcrowding effects and Allee or undercrowding effects), and both demographic and environmental stochasticity. The form of density dependence (e.g., logistic, Allee, etc.), the mean and standard deviation of growth rates, and the carrying capacities of each population may be different. Spatial structure is introduced in two ways: (1) the growth rates are correlated among populations, and (2) at each time period, a specified proportion of individuals migrate from each population to others. Both correlations and migrations are specified as $n \times n$ matrices, where n is the number of populations. It is also possible to specify the correlation coefficients and migration rates as functions of distance: both the similarity of environments and the rate of migration are likely to decrease as the distance between two populations increases.

Conclusion

The extinction risk of a species is determined by a multitude of factors, such as life history characteristics, effects of environmental and demographic stochasticity, number and geographic configuration of its local populations, and environmental correlation and migration among populations. Assessment of ecological risks must be based on as much information as possible on the biology of the endangered species and on the characteristics of its metapopulation. Similarly, decisions about the design of nature reserves and management options such as stocking, translocation and reintroduction of individuals as well as assessment of human impact will benefit from the utilization of such information.

The single and multiple population models discussed in this paper allow the incorporation of species-specific data in models of age- or stage-structured populations and spatially-structured metapopulations. The probabilistic framework of these models allow estimation of extinction risks in addition to prediction of future population trajectories. They can be used to predict probabilities of population decline to specified levels under different conditions or management practices. Such comparative studies and sensitivity analyses provide a realistic methodology for making conservation decisions based on biological data.

Zusammenfassung

Es werden drei mathematische Modelle zur Bewertung des Aussterberisikos vorgestellt. Das erste ist ein Monte-Carlo-Modell, das das Wachstum einer Population mit einer Altersstruktur unter einer umweltbedingten und demographischen Stochastizität beschreibt. Die Parameter sind der Mittelwert, die Variation, die Verteilung und Korrelationen von Eigenschaften wie Überleben, Fruchtbarkeit und Migration, sowie Dichteabhängigkeit der Vermehrung. Die Wahrscheinlichkeiten des Populationsrückgangs und des Aussterbens werden berechnet und stellen eine Möglichkeit dar, das ökologische Risiko zu bewerten. Die beiden anderen Modelle beschreiben die Dynamik von Systemen mit mehreren Populationen (Metapopulationen), wobei angenommen wird, daß deren räumliche Struktur und die Migrationsraten zwischen Populationen bekannt sind. Das eine dieser Modelle von Metapopulationen basiert auf der Besetzung von Habitatausschnitten, das andere auf der Populationsdynamik innerhalb lokaler Populatonen. Diese Modelle bewerten sowohl das Riskiko des Aussterbens der ganzen Metapopulation als auch das lokaler Populationen.

Es wurden drei Computerprogramme (die RAMAS-Bibliothek) für das Erstellen von Modellen für einzelne oder multiple Populationen beschrieben. Die Programme werden benutzt, um Modelle zu erstellen, die alters-, stadien- oder räumlich strukturiert sind. Sie liefern eine Übersicht bezüglich der Individuenzahl in Abhänigkeit der Zeit und die Wahrscheinlichkeit, daß die Populationsgröße unter einen bestimmten Schwellwert fällt.

Acknowledgments

We thank SCOTT FERSON, GEOFF JACQUEZ and JEFF MILLSTEIN for their assistance. RAMAS computer programs are available through Exeter Software (1-516-689-7838, fax: 1-516-751-3435).

References

AKÇAKAYA, H. R. and S. FERSON. 1990. RAMAS/space User Manual: Spatially-structured Population Models for Conservation Biology. Exeter Software, New York.

BOYCE, M. S. 1977. Population growth with stochastic fluctuations in the life table. Theoretical Population Biology 12:366-373.

BURGMAN, M. A., H. R. AKÇAKAYA and S. S. LOEW. 1988. The use of extinction models in species conservation. Biological Conservation 43:9-25.

DOMBROVSKII, YU. A. and TYUTYUNOV, YU. V. 1987. Habitat structure, mobility of individuals and persistence of populations. Journal of General Biology 4:493-498 [in Russian].

FERSON, S. 1990. RAMAS/stage User Manual: Generalized Stage Modeling for Population Dynamics. Exeter Software, Setauket, New York.

FERSON, S., L. R. GINZBURG and A. SILVERS. 1989. Extreme event risk analysis for age-structured populations. Ecological Modelling 47:175-187.

FERSON, S., R. AKÇAKAYA, L. GINZBURG and M. KRAUSE. 1990. Application of RAMAS to the analysis of ecological risk: examples from two species of fish. Electric Power Research Institute (in press).

FERSON, S. and H. R. AKÇAKAYA. 1990. RAMAS/age User Manual: Modeling Fluctuations in Age-structured Populations. Exeter Software, New York.

GILPIN, M. E. 1988. A comment on Quinn and Hastings: extinction in subdivided habitats. Conservation Biology 2:290-292.

GINZBURG, L. R., K. JOHNSON, A. PUGLIESE and J. GLADDEN. 1984. Ecological risk assessment based on stochastic age-structured models of population growth. Special Technical Testing Publication 845:31-45.

GINZBURG, L. R., L. B. SLOBODKIN, K. JOHNSON and A. G. BINDMAN. 1982. Quasiextinction probabilities as a measure of impact on population growth. Risk Analysis 2:171-181.

GINZBURG, L. R., S. FERSON and H. R. AKÇAKAYA. 1990. Reconstructibility of density dependence and the conservative assessment of extinction risks. Conservation Biology 4:63-70.

HANSKI, I. 1989. Metapopulation dynamics: does it help to have more of the same? Trends in Ecology and Evolution 4:113-114.

HARCOURT, A. H., D. FOSSEY and J. SABATER-PI. 1981. Demography of Gorilla gorilla. Journal of Zoology 195:215-233.

HARRISON, S. and J. F. QUINN. 1989. Correlated environments and the persistence of metapopulations. Oikos 56:293-298.

LANDE, R. and S. H. ORZACK 1988. Extinction dynamics of age-structured populations in a fluctuating environment. Proceedings of the National Academy of Sciences 85:7418-7421.

LEIGH, E. G. JR. 1981. The average lifetime of a population in a varying environment. Journal of Theoretical Biology 90:213-239.

LEVINS, R. 1969. The effects of random variation of different types on population growth. Proceedings of the National Academy of Sciences 62:1061-1065.

LEVINS, R. 1970. Extinction. In: Some mathematical questions in biology. M. Gerstenhaber (ed.) American Mathematical Society, Providence, Rhode Island.

LEWONTIN, R. C. and D. COHEN. 1969. On population growth in a randomly fluctuating environment. Proceedings of the National Academy of Sciences 62:1056-1060.

MACARTHUR, R. H. and E. O. WILSON. 1967. The theory of island biogeography. Princeton University Press, Princeton, New Jersey.

MARGULES, C., A. J. HIGGS and R. W. RAFE. 1982. Modern biogeographic theory: are there any lessons for reserve design? Biological Conservation 24:115-128.

MAY, R. M. 1973. Stability in randomly fluctuating versus deterministic environments. American Naturalist 107:621-650.

PIMM, S. L., H. L. JONES and J. DIAMOND. 1988. On the risk of extinction. American Naturalist 132:757-785.

QUINN, J. F. and A. HASTINGS. 1987. Extinction in subdivided habitats. Conservation Biology 1:198-208.

ROUGHGARDEN, J. 1975. A simple model for population dynamics in stochastic environments. American Naturalist 109:713-736.

SHAFFER, M. L. 1983. Determining minimum viable population sizes for the grizzly bear. International Conference on Bear Research and Management 5:133-139.

SHAFFER, M. L. and F. B. SAMSON. 1985. Population size and extinction: a note on determining critical population sizes. American Naturalist 125:144-152.

SIMBERLOFF, D. and L. G. ABELE. 1982. Refuge design and island biogeographic theory: effects of fragmentation. American Naturalist 120:41-50.

TULJAPURKAR, S. D. and S. H. ORZACK. 1980. Population dynamics in variable environments. I. Long-run growth rates and extinction. Theoretical Population Biology 18:314-342.

WEBBER, A. W. and A. VEDDER. 1983. Population dynamics of the Virunga Gorillas: 1959-1978. Biological Conservation 26:341-366.

Genetic and Phenotypic Variation in Relation to Population Size in Two Plant Species: *Salvia pratensis* and *Scabiosa columbaria*

R. Bijlsma [1], **N.J. Ouborg** [2] and **R. van Treuren** [1]

[1] Department of Genetics, University of Groningen, Kerklaan 30, 9751 NN HAREN,

[2] Department of Plant Ecology, Institute for Ecological Research, P.O. Box 40, 6666 ZG HETEREN, The Netherlands.

Abstract

Due to human activities many populations have become small, fragmented and isolated and consequently more sensitive to genetic drift and inbreeding, resulting in loss of genetic variation and fixation of deleterious alleles. This process often involves a decrease in viability (inbreeding depression) and therefore may significantly affect the probability of extinction of populations. To assess the relevance of these predictions, the amount of genetic variation was measured in relation to population size for two plant species, *Salvia pratensis* and *Scabiosa columbaria*. Both species are endangered in The Netherlands and show a considerable decline in number of populations during the last 30 years. For allozymes significant correlations were observed between population size and both the proportion of polymorphic loci and the mean observed number of alleles, the large populations being more variable than the small populations. In addition, substantial genetic differentiation was observed between populations, the differentiation being more extensive among small populations than among large populations. Concerning morphological characters also a positive relationship between the total amount of phenotypic variation and population size was observed. The results suggest that, predominantly due to genetic drift, small populations are depauperate in genetic variation. Although small populations, at least for *Salvia*, showed some indications of decreased viability, much more research is needed to decide whether these low levels of genetic variation influence the probability of extinction significantly.

Introduction

It is wellknown that in the course of evolution many species have become extinct. Based on fossil records it has been estimated that of all species that ever evolved on earth 99% has vanished again. So extinction of species *per se* can by no means be regarded as an "unnatural" phenomenon and it is therefore only logical that it will also occur today. However, due to the devastating activities of man, we have recently been confronted with an increase in extinctions or even mass extinctions that is unprecedented and could therefore be regarded as "unnatural". In some cases the extermination has been directly by hunting and killing of the species. A salient example in this respect is the Steller's sea cow (*Rhytina gigas*) of which the last individual was killed in 1768, a mere 27 years after its discovery. Most often, however, human induced extinctions are caused through habitat alteration and habitat destruction, threatening the existence of many plant and animal species (FRANKEL & SOULÉ 1981, SOULÉ 1983). Apart from direct destruction of natural habitats for agricultural development, industrialization and urbanization, another important factor is the introduction of exotics that become predators or competitors of the native species. This factor has been shown to be very important in isolated ecosystems; especially island systems seem to be very vulnerable in this respect. Destruction of natural habitats, however, is by far the most important cause of species extinction at the moment. For instance the tropical rain forest, thought to be the most species rich habitat, is destroyed at a very fast rate, together with the species it sustains. If the destructive activities of man proceeds in this way, it is estimated that 10-20% of all species will be threatened to become extinct in the next few decades. Although less is known about extinctions of plant species as compared to animals, a conservative estimate by the IUCN'S Threatened Plants Unit (1983) predicts that around 10% of the world's flowering plants are endangered within this period.

Habitat Fragmentation and Genetic Aspects

The development of natural habitats for agricultural use etc., not only causes reduction in size of the habitat, but also often results in fragmentation of the remaining area, leaving isolated patches of habitat in an otherwise uninhabitable "desert" (see examples by CURTIS, 1956; WILCOVE *et al.*, 1986). As each of the remaining fragments can sustain only a small population, the probability of extinction through random demographic and environmental forces does increase (GOODMAN, 1987; SHAFFER,1987). If population sizes become small, also genetic problems may arise. Firstly, under these circumstances population genetic theory predicts loss of genetic variation due to random genetic drift. The rate of loss strongly depends on the (effective) population size and increases when population size becomes smaller. Not only deleterious or neutral alleles will be lost in this way but advantageous alleles as well. This loss of variation may significantly limit the adaptability of the population in changing environments in future generations. Secondly, in small populations individuals become related to one another and consequently we may face the problem of inbreeding. As inbreeding is most often associated with a significant loss in viability, due to fixation of deleterious alleles (inbreeding depression), this may result in a decrease in fitness (CHARLESWORTH & CHARLESWORTH, 1987).

The consequences of habitat fragmentation and the above mentioned processes strongly depend on the amount of migration or gene flow that is still possible between isolated fragments. If gene flow is sufficiently large, we can still consider all fragments as one, though subdivided, population. In this so called metapopulation structure, local extinction of subpopulations may occur but these can be undone by recolonization from still existing areas (see e.g. GILPIN, 1987). Though loss of genetic variation may occur in individual subpopulations, different alleles will be lost in different subpopulations, so that genetic variation is still preserved in the metapopulation. Migration at a sufficient rate will also prevent the problem of inbreeding. However, if gene flow between the fragments is absent or very restricted, which might be the case for many sessile plant species, we have to regard the individual fragments as isolated populations of small size. Consequently each individual population will be subject to loss of genetic variation and inbreeding depression.

Although the negative effects of loss of genetic variation and inbreeding depression are well known from artificial selection and husbandry (for many examples see FALCONER, 1980) empirical evidence of the action of these processes in nature is still scarce and most data concern large mammals. A well documented example of the relation between homozygosity and fitness has been given by O'Brien and coworkers. They showed that the African cheetah is nearly homozygous for more than 200 protein loci (O'BRIEN et al., 1983; 1987). By a number of reciprocal skin tissue transplants they showed that even the major histocompatibility complex, the most variable locus in vertebrates, was invariable (O'BRIEN et al., 1985). This lack of genetic variation, most probably caused by bottlenecks in population size, was accompanied by a high incidence of abnormal spermatozoa causing low fertility, a high juvenile mortality and an increased sensitivity to diseases. This strongly suggests that the cheetah really suffers from inbreeding depression. The data on inbreeding depression available so far (CHARLESWORTH & CHARLESWORTH,1987) indicate that its impact can be very large in both animal and plant species, and it is argued that the effect of inbreeding is rather under than overestimated.

The importance of genetic variation at the individual level has been demonstrated by SCHAAL & LEVIN (1976) for the perennial plant species *Liatris cylindracea*. This species is longlived, up to 44 years, and has the advantage that the age of an individual can be determined from its corm. Schaal and Levin showed that the relative contribution of heterozygotes (with respect to allozyme loci) in the population increased with increasing age of the individual plants, indicating that the more heterozygous plants live longer than homozygotes. In addition, a positive relationship was observed between individual heterozygosity and fecundity, rate of development and age at sexual maturity. This indicates that heterozygosity is being selected for in *Liatris*. A positive correlation between growth rate and heterozygosity has been observed now for many organisms (for a review see MITTON & GRANT, 1984).

Plants

Plants generally differ significantly from animals for a number of important characteristics which bear relation to the above mentioned extinction problems. The most significant of these are:

1) Most plants are sessile and except for wind pollinated species gene flow is generally restricted. Outcrossing plants, therefore, may suffer significantly more from fragmentation than animals. On the other hand many hermaphrodite plants are predominantly selfing and consequently accustomed to a considerable degree of inbreeding. Moreover, because of very restricted pollenflow, many large plant populations in fact may have become subdivided in small breeding units (neighborhood sizes) in which plants have become consanguineous, resulting also in a certain degree of inbreeding. Therefore, in these cases plants may be less sensitive to decreased population size.

2) Plants, once established, have to be adapted to their local environment because they cannot move to a more favorable spot like animals. They can adapt both phenotypically, as plants often show a considerable plasticity, and genotypically, by selection. The latter can lead to local adapted gene complexes that are specific for the local habitat. Gene flow between habitats in such cases will cause hybridization and consequently a breakdown of the adapted complex by recombination, followed by a decreased fitness of the offspring. This phenomenon is called outbreeding depression (TEMPLETON, 1986). Indications of the existence of this phenomenon in nature is shown by PRICE & WASER (1979) and WASER & PRICE (1983). They observed for *Delphinium nelsonii* optimal outcrossing in relation to the geographical distance between the parental plants: parents close to each other may be consanguineous and their offspring the result of inbreeding while parents far from each other may be genetically dissimilar and their offspring can suffer from outbreeding depression. Optimal offspring fitness will therefore be achieved when the parents are at an intermediate distance which ensures that they are no relatives but still are adapted to the same local environment. The existence of outbreeding depression may be of significant importance for the management of plant populations and indicates that artificial gene flow might not be favorable in every case.

3) Plants do show a wide range of breeding systems, from asexual reproduction to pure outcrossing. Even within populations of a species variation may be present e.g. both hermaphrodite plants and male sterile plants (RICHARDS, 1986). The breeding system has a significant impact on the distribution and maintenance of genetic variation within populations, both by influencing the effective population size and the amount of inbreeding versus outcrossing.

Experimental Results

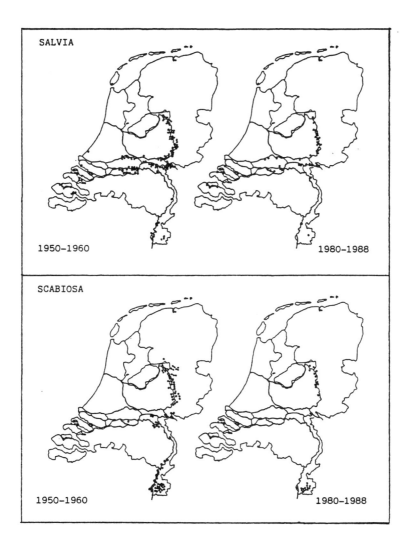

Fig. 1:   Approximate number of populations of *Salvia* (top) and *Scabiosa* (bottom) recorded in The Netherlands during the time periods 1950-60 (left) and 1980-88 (right). Each dot represents a site the species was recorded to exist.

Empirical data concerning the importance of genetic variability in relation to the extinction process of species are still scarce, and in case of plants, almost nonexistent. Therefore, we have started a research project to assess the role of these genetic aspects for two plant species. This project is a joint venture of the Department of Genetics of the University of Groningen at Haren and the Institute for Ecological Research at Heteren. The project was started 2 years ago and the results we have obtained so far are presented here.

For our research we have selected the plant species: *Salvia pratensis* (meadow sage) and *Scabiosa columbaria* (small scabious). Both species are perennial, predominantly outcrossing and both are gynodioecious, which means that both hermaphrodite plants and female plants are present in the population. *Salvia* is a labiate that can be found in calcareous grasslands on riverdunes and river dikes and, in The Netherlands, occurs mostly along the big rivers (see Fig. 1) *Scabiosa*, a Dipsacacea, can be found in dry grassy sites on calcareous soils: its distribution is not only along the rivers but it is also found in the limestone district in the south of The Netherlands. Both species show a sharp decline in number of populations during the last decades, as can be seen in Fig. 1. This figure shows for both species the number of populations recorded during the fifties, compared to the situation in the eighties. A decrease of more than 50% is apparent from the comparison. Fortunately still a sufficient number of large and small populations are present for research. Our researches focus on the following aspects and comparisons will be made between large and small populations concerning these aspects.

1)   The amount of genetic variation both for allozymes and quantitative characters.

2)   Demographic measures e.g. effective population size, reproductive capacity etc.

3)   The amount of gene flow and outcrossing rates.

4)   The degree of inbreeding depression in populations (if present) and the sensitivity of the species to inbreeding.

Integration of the results with respect to these points will hopefully give sufficient information about the key factors that might be important in relation to the extinction of these two species and by extrapolation also for other plant species. Up till now most data we obtained concern the amount of genetic variation in small and large populations.

Allozymes

By means of electrophoresis the amount of allozyme variation was measured for both species (VAN TREUREN *et al.*, 1991). *Salvia* was screened for 13 loci of which, combined over all populations, 11 (=84.6%) were polymorphic. For *Scabiosa* these data were 12 and 10 (=83.3%), respectively. The proportion of polymorphic loci per population varied between 0.25 and 0.60 for both species, which falls well within the range generally observed for plants (NEVO *et al.*, 1984). More important, however, is that for both species a significant positive correlation exists between the proportion of polymorphic loci and population size (Fig. 2A).

For *Salvia* the correlation is quite convincing, a bit less so for *Scabiosa*, mainly due to the lack of intermediate population sizes. Correlations are also observed between the mean observed number of alleles, a measure of the number of different genotypes that are possible in the population, and population size (Fig. 2B). Though one would expect a correlation to exist between population size and gene diversity (mean expected heterozygosity, computed

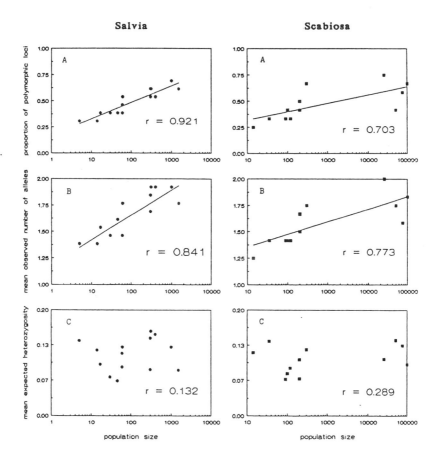

Fig. 2: Relation between population size (log scale) and different measures for the level of genetic variation within populations for both **Salvia** (left) and **Scabiosa** (right). The measures are: proportion of polymorphic loci (A); mean observed number of alleles (B); mean expected heterozygosity (C). In each case the linear correlation coefficient, r, is also given together with the regression line when significant.

according to NEI, 1987) also, no such significant correlation was observed (Fig. 2C). This was in both species mainly due to the smallest populations showing relatively far too high values. The reason for this discrepancy is not clear at the moment, but might be partly due to the fact that we used the observed number of individuals and not effective population sizes for our computations, because the latter have not yet been determined. A comparable situation has been observed by (VARVIO-AHO, (1981) for the Finnish waterstrider. But it also might indicate that population size is not the only factor determining the level of genetic variation. To clarify this question we have to assess other demographic and life history parameters and correlate these to population size.

When populations become isolated we might expect them to diverge genetically. This was analyzed using Nei's statistics of gene diversity (NEI, 1987). The proportion of the total genetic diversity found among populations, the coefficient of gene differentiation, G_{ST}, was calculated for both species and was found to be 0.156 and 0.175 for *Salvia* and *Scabiosa*, respectively (Table 1). These G_{ST}-values are substantially higher than the values on the average observed for predominantly outcrossing or longlived species, 0.118 and 0.077 respectively, but, on the other hand, lower than those observed for species showing mixed mating, 0.243 (LOVELESS & HAMRICK, 1984). If genetic drift is the main cause for the diversity among populations, we would expect differentiation among small populations to be more extensive than among large populations. Therefore, for both species the populations were grouped into two size classes and G_{ST}-values among small and among large populations were calculated. Table 1 shows that the values observed for small populations are indeed considerably higher than observed for large populations. This differentiation was especially marked for *Scabiosa*. Substantial differentiation was not only observed between geographically isolated populations, but also for neighboring populations growing in apparently similar environments. This suggests that gene flow between populations is restricted and that random processes might be predominant in small populations of both species.

Tab. 1: The coefficient of gene differentiation, G_{ST}, calculated for all populations combined and calculated for the small and large populations separately. The number of populations is given in brackets.

	all populations	small populations	large populations
Salvia	0.156 (n=14)	0.181 (n=8)	0.115 (n=6)
Scabiosa	0.175 (n=12)	0.236 (n=5)	0.101 (n=7)

Morphological Characters

In addition to allozymes the level of morphological variation within populations of both species was studied also (OUBORG *et al.*, 1991). For this purpose seeds were collected from five *Salvia* and seven *Scabiosa* populations. Of each population 6-7 families, usually consisting of 5 to 10 offspring per mother plant, were grown under uniform conditions in a glasshouse. These were used to measure a number of morphological characters including characters of the seedling, juvenile and adult life stage and, for *Scabiosa* only, reproductive characters. As the seeds were sampled directly from the field, the offspring within a family may be either the result of selfing, a fullsib outcross or a halfsib outcross, which makes a straightforward quantitative genetic analysis difficult. However, an analysis of variance revealed significant differences both between populations and between families within populations for many of these characters. This indicates that at least part of the observed phenotypic variation can be ascribed to genetic variation. In order to combine the data for all characters simultaneously the phenotypic variation was expressed as coefficient of variation (CV) and for each character the populations, within each species, were ranked according to their CV-value. Next, for each population the mean rank over all characters was calculated. In figure 3 this mean rank of CV-value is plotted against the populations ranked by size for both *Salvia* and *Scabiosa*. It is clear that small populations do show less variation than large populations. The correlation between the population size rank and the mean rank of CV-value was significant for Salvia ($r = 0.892$; $P < 0.05$) and at the border of significance for *Scabiosa* ($r = 0.730$; $0.05 < P < 0.10$). It is noteworthy that for both species the smallest population seems to show a relatively high level of variation, a situation comparable to the one observed in the allozyme study with respect to the mean expected heterozygosity.

Although there are strong indications that the phenotypic variation is at least partly due to genetic variation, it still has to be determined more precisely to what extent the observed morphological variation is genetically determined. Nevertheless, the results so far indicate that for both species, also with respect to these morphological characters, small populations do show significantly less variation than large populations.

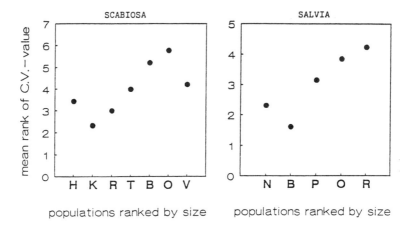

Fig. 3:   Relation between population size and the amount of morphological variation within populations for **Scabiosa** (left) and **Salvia** (right). The letters denote the different populations and are ranked according to ascending size.
For further explanation see text.

Discussion and Conclusions

Notwithstanding the fact that many life history characteristics other than population size also do affect the amount and the distribution of genetic variation significantly (LOVELESS & HAMRICK, 1984), we found for both *Salvia* and *Scabiosa* a significant correlation between the size of a population and its level of polymorphism with respect to allozymes. Comparable results were reported for other plant species (MORAN & HOPPER,1983; KARRON, 1987). The relationship between population size and polymorphism suggests that the smaller populations were less variable, although the correlation between size and gene diversity was not found to be significant. Moreover, the level of morphological variation seemed also to be positively correlated with population size. It has been argued that variation at allozyme loci often can be considered to be selectively neutral or nearly neutral and therefore is mainly affected by random processes, while many morphological characters most often are of adaptive significance and consequently rather affected by natural selection (TURNER *et al.*, 1979; WOLFF, 1988). The observation that both allozymic and morphological variation show a similar relation with population size, indicates, therefore, that genetic drift might have been largely responsible for the loss of variation in the smaller populations. ALLENDORF *et al.* (1982) and LAGERCRANTZ & RYMAN (1990) observed concordance between allozymic and morphological variation in a number of coniferous tree species, that was also most probably accounted for by genetic drift and/or bottleneck effects.

The lower level of genetic variation observed in the small populations, however, does not necessarily imply that these populations also should show a decreased viability. In contrast to what has been observed for the cheetah (see above), other species are known that show little or no genetic variation but nevertheless show a good viability. Such an example is the northern elephant seal (*Mirounga angustirostris*) which supposedly has gone through a severe bottleneck within the last century and showed no electrophoretically detectable variation at all, but nevertheless showed a rapid increase in population size thereafter (BONNELL & SELANDER, 1974). Moreover, many obligatory or predominantly inbreeding plants also seem to suffer no ill effects and are very successful, although CHARLESWORTH & CHARLESWORTH (1987) argue that even in these cases outcrossing might result in a significant fitness gain. In order to relate genetic depauperization to inbreeding depression and extinction, one has to establish that small populations actually show decreased fitness compared to large populations. Therefore we are currently assessing the effects of inbreeding on a number of fitness related characters directly by making different crosses (selfings, outcrosses both within and between populations). Also we are comparing small and large populations with respect to these characters. There are indirect indications that fitness differences might indeed be present: a Discriminant Analysis of the morphological data revealed for *Salvia* considerable differences between populations in growth related characters such as number of leaves and number of rosettes. The small populations showed a lower "growth rate" than the large populations, indicating that small populations might have a decreased competitive ability compared to large populations. On the other hand, for *Scabiosa* no such relationship between population size and fitness related characters was observed.

Our results so far are in agreement with the expectation that small populations show decreased levels of genetic variation, most probably caused by random processes. Whether this affects fitness and enhances the chance of extinction of these populations still has to be decided.

Zusammenfassung

Infolge der durch den Menschen verursachten Umgestaltung ihres Lebensraums wurden viele Populationen verkleinert, fragmentiert und isoliert. Solche Populationen sind meist sensitiver gegenüber genetischer Drift und Inzucht, zweier Faktoren, die einen Verlust von genetischer Variabilität zur Folge haben und zur Fixierung von nachteiligen Allelen führen können. Der damit einhergehende Verlust an Überlebensfähigkeit (Inzuchtdepression) kann die Wahrscheinlichkeit des Aussterbens von Populationen signifikant beeinflussen. Um die Relevanz dieser Vorhersagen abzuschätzen, wurde das Ausmaß der genetischen Variabilität im Verhältnis zur Populationsgröße an zwei Pflanzenarten, *Salvia pratensis* und *Scabiosa columbaria*, gemessen. Beide Arten sind in den Niederlanden gefährdet und zeigen einen beträchtlichen Rückgang der Anzahl der Populationen während der letzten 30 Jahre.

Bei den Allozymen wurden signifikante Korrelationen zwischen der Populationsgröße und dem Anteil der polymorphen Loci sowie der mittleren Anzahl der beobachteten Allele gefunden. Die größeren Populationen waren variabler als die kleinen Populationen.

Zusätzlich wurde eine deutliche genetische Differenzierung zwischen den Populationen beobachtet. Die Differenzierungen waren ausgeprägter zwischen den kleineren Populationen als zwischen den größeren Populationen. Hinsichtlich der morphologischen Eigenschaften wurde ebenfalls eine positive Beziehung zwischen phänotypischer Variabilität und der Populationsgröße beobachtet.

Die Ergebnisse lassen vermuten, daß kleine Populationen hauptsächlich aufgrund der genetischen Drift in ihrer genetischen Variabilität verarmt sind. Obwohl kleine Populationen, zumindest bei *Salvia*, einige Hinweise auf eine verringerte Überlebensfähigkeit zeigten, ist mehr Forschung erforderlich um zu entscheiden, ob dieses geringe Ausmaß an genetischer Variabilität die Wahrscheinlichkeit signifikant beeinflußt, daß eine Population ausstirbt.

Acknowledgments

We thank Dr. W. Van Delden for critically reviewing the manuscript. This research was partly supported by the Ministry of Agriculture, Nature Management and Fishery.

References

ALLENDORF, F.W., K.L. KNUDSEN & G.M. BLAKE (1982): Frequencies of null alleles at enzyme loci in natural populations of ponderosa and red pine. Genetics 100: 497-504.

BONNELL, M.L. & R.K. SELANDER (1974): Elephant seals: genetic– variation and near extinction. Science 184: 908-909.

CHARLESWORTH, D & B. CHARLESWORTH (1987): Inbreeding depression and its evolutionary consequences. Ann. Rev. Ecol. Syst. 18:237-268.

CURTIS, J.T. (1956): The modification of midlatitude grasslands and forests by man. In: W.L. THOMAS (ed) Man's Role in Changing the Face of Earth. University of Chicago Press, Chicago.

FALCONER, D.S. (1980): Introduction to Quantitative Genetics 2nd ed. Longman, London.

FRANKEL, O.H. & M.E. SOULÉ (1981): Conservation and Evolution. Cambridge University Press, Cambridge.

GILPIN, M.E. (1987): Spatial structure and population vulnerability. In: M.E. SOULÉ (ed) Viable Populations for Conservation. Cambridge University Press, Cambridge.

GOODMAN, D. (1987): The demography of chance extinction. In: M.E. SOULÉ (ed) Viable Populations for Conservation. Cambridge University Press, Cambridge

IUCN Threatened Plant Unit (1983): List of rare, threatened and endemic plants in Europe 1982. Ed. Council of Europe Natureand Environment Series 27, Strassbourg.

KARRON, J.D. (1987): A comparison of levels of genetic polymorphism and self compatibility in geographically restricted and widespread plant congeners. Evol. Ecol. 1: 47-58.

LAGERCRANTZ, U. & N. RYMAN (1990): Genetic structure of Norway spruce (Picea abies): Concordance of morphological and allozymic variation. Evolution 44: 38-53.

LOVELESS, M.D. & J.L. HAMRICK (1984): Ecological determinants of genetic structure in plant populations. Ann. Rev. Ecol.Syst. 15: 65-95.

MITTON, J.B. & M.C. GRANT (1984): Associations among protein heterozygosity, growth rate, and developmental homeostasis. Ann. Rev. Ecol. Syst. 15: 479-499.

MORAN, G.F. & S.D. HOPPER (1983): Genetic diversity and the insular population structure of the rare granite rockspecies, *Eucalyptus caesia*. Aust. J. Bot. 31: 161-172.

NEI, M. (1987): Molecular Evolutionary Genetics. Columbia Univerisity Press, New York.

NEVO, E., A. BEILES & R. BEN-SHLOMO (1984): The evolutionary significance of genetic diversity: Ecological, demographic and life history correlates. In: G.S. MANI (ed) Evolutionary Dynamics of Genetic Diversity, pp 13-213. Springer-Verlag, Berlin.

O'BRIEN, S.J., M.E. ROELKE, L. MARKER, A. NEWMAN, C.A. WINKLER, D. MELTZER, L. COLLY, J.F. EVERMANN, M. BUSH & D.E. WILDT (1985): Genetic Basis for species vulnerability in the cheetah. Science 227: 1428-1434.

O'BRIEN, S.J., D.E. WILDT, D. GOLDMAN, C.R. MERRIL & M. BUSH (1983): The cheetah is depauperate in genetic variation. Science 221: 459-462.

O'BRIEN, S.J., D.E. WILDT, M. BUSH, T.M. CARO, C. FITZGIBBON, I. AGGUNDEY & R.E. LEAKEY. (1987): East African cheetahs:Evidence for two population bottlenecks? Proc. Natl. Acad.Sci. USA 84: 508-511.

OUBORG, N.J., R. VAN TREUREN & J.M.M. VAN DAMME (1991): The significance of genetic erosion in the process of extinction II. Morphological variation and fitness components in populations of varying size of *Salvia pratensis* L. and *Scabiosa columbaria* L.. Oecologia, in press.

PRICE, M.V. & N.M. WASER (1979): Pollen dispersal and optimal outcrossing in *Delphinium nelsonii*. Nature 227: 294-297.

RICHARDS. A.J. (1986): Plant Breeding Systems. ALLEN and UNWIN, London. SCHAAL, B.A. and D.A. LEVIN (1976) The demographic genetics of *Liatris cylindracea* Michx. (Compositea). Amer. Natur. 110:191-206.

SHAFFER, M. (1987): Minimum viable populations: coping with uncertainty. In: M.E. SOULÉ (ed) Viable Populations for Conservation. Cambridge University Press, Cambridge.

SOULÉ, M.E. (1983): What do we really know about extinction? In: C.M. SCHONEWALD-COX, S.M. CHAMBERS, B. MCBRYDE and W.L. THOMAS (eds) Genetics and Conservation: A reference formanaging wild animal and plant populations. Benjamin Cummings Publ. Co., Menlo Park, Calif.

TEMPLETON, A.R. (1986): Coadaptation and outbreeding depression. In: M.E. SOULÉ (ed) Conservation Biology: Science of Diversity and Scarcity. Sinauer Associates, Sunderland, Mass.

TURNER, J.R.G., M.S. JOHNSON & W.F. EANES (1979): Contrasted modes of evolution in the same genome: allozymes and adaptive changes in Heliconius. Proc. Natl. Acad. Sci. USA 76:1924-1928.

VAN TREUREN, R., R. BIJLSMA, W. VAN DELDEN & N.J. OUBORG (1991): The significance of genetic erosion in the process ofextinction. 1. Genetic differentiation in *Salvia pratensis* and *Scabiosa columbaria* in relation to population size. Heredity, in press.

VARVIO-AHO, S.L. (1981): The effects of ecological differences on the amount of enzyme gene variation in Finnish waterstrider (Gerris) species. Hereditas 94: 35-39.

WASER, N.M. & M.V. PRICE (1983): Optimal and actual outcrossing in plants, and the nature of plant pollinator interaction.In: C.E. JONES and R.J. LITTLE (eds) Handbook of Experimental Pollination Biology. Scientific and Academic Editions, New York.

WILCOVE, D., C. MCLELLAN & A. DOBSON (1986): Habitat fragmentation in the temperate zone pp 237-56. In: M.E. SOULÉ (ed) Conservation Biology: Science of Diversity and Scarcity. Sinauer Associates, Sunderland, Mass.

WOLFF, K. (1988): Natural selection in Plantago species: a genetical analysis of ecologically relevant morphological variability. Ph.D. Thesis. University of Groningen.

Plasticity in Life History Traits of the Freshwater Pearl Mussel - Consequences for the Danger of Extinction and for Conservation Measures

G. Bauer, Department of Animal Ecology I, University of Bayreuth, D-8580 Bayreuth, Germany

Abstract

In Central Europe, the freshwater pearl mussel (*Margaritifera margaritifera*) is threatened with extinction. It is shown how the danger of extinction is influenced by the plasticity of life history traits. The pearl mussel's life history strategy has been selected for a high lifetime fertility, which is attained by combining high life expectancy (= high number of reproductive periods) and high fertility. However, these two traits exhibit a considerable plasticity, which depends on the individual growth constant (i.e. the rate at which the asymptotic size is approached). The growth constant is influenced by a number of environmental factors: for example it increases with increasing temperature. An increased growth constant (= accelerated growth) leads to a reduction of life span, maximum size and fertility. These relationships reduce the lifetime fertility of individuals growing at a high rate. Populations consisting of such individuals should exhibit a high sensitivity to threats for two reasons: the growth rate of the population is low and the time until extinction is shortened due to the low individual life span. This hypothesis is confirmed by the population trend in different areas of the FRG. The results are discussed with respect to conservation strategies.

Introduction

Integrated Pest Management and species conservation do pursue opposite goals with respect to the target organism. However, both purposes require nearly the same detailed knowledge about the population ecology of the target organism. Investigations must therefore comprise the autecology as well as demographic processes in relation to extrinsic and intrinsic factors (BURGMANN *et al.* 1988). Ideally, such research programs should yield knowledge about minimum viable populations (SCHAFFER 1981), about weak points in the life cycle and about the most important factors affecting density. In the case of species conservation these data are then used to increase or stabilize the density of the target organism.

However, such investigations are costly and take a lot of time and can, therefore, usually be conducted only at few localities. This raises the question whether or not the results can be applied to other populations of the same species. The problem does not only exist because the threats may be locally different. Populations may also differ with respect to their fitness due to plasticity in life history traits. On principle, all life history traits, like body size, longevity, fecundity, etc. may respond to a variable environment. However, usually some traits are fixed (are resistant to change) whereas others are highly plastic (BERVEN & GILL 1983). The fixed traits, the degree of plasticity and compensating mechanisms are not predictable but must be investigated for each species.

Plasticity may be adaptive or nonadaptive. Adaptive plasticity is an optimal response to a specific set of environmental stimuli which has a genetic basis, whereas nonadaptive plasticity is a nonspecific, passive variation of the phenotype due to physiological constraints of the environment (SMITH-GILL 1983). Particularly nonadaptive plasticity may be an important factor for the status of populations and for conservation strategies, but it is usually neglected in this context. The interplay of plasticity and threats as well as the consequences with respect to conservation shall therefore be described for the freshwater pearl mussel (*Margaritifera margaritifera L.*).

Biology of the freshwater pearl mussel

Freshwater pearl mussels inhabit exclusively running waters which are poor in nutrients. In late summer the females release glochidia. They can only develop into young mussels if they are able to attach to the gills of a suitable host fish, which they reach passively in the ventilating current of the fish. The only important host in Central Europe is brown trout (*Salmo trutta*, BAUER 1987a). The glochidia attach to the gills of the host and live as parasites for some weeks up to nine months. During this period they grow from 0.07 mm to 0.4 mm. Having completed metamorphosis the young mussels leave the host and presumably live for 3-5 years burrowed in the river bottom before they appear at the surface and begin their adult life.

Life history strategy

A very important factor influencing the evolution of the life history strategy are extremely high mortality rates of the glochidia. Out of one million glochidia only less than ten are lucky to be inhaled by a susceptible host (BAUER 1989, YOUNG & WILLIAMS 1984). To investigate the density relationships of this survival rate, the densities of parasitizing glochidia were plotted against the numbers of adult mussels present in the rivers (Fig. 1). The slope of the regression cannot be distinguished from b = 1, suggesting that in all populations the survival rate is the same (SOLOMON 1968). Therefore this relationship must be considered as density-independent in the investigated range of densities.

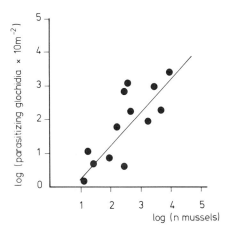

Fig. 1: Relationship between number of adult pearl mussels present in the river and density of parasitizing glochidia ($y = -0.5 + 0.92x$; $r = 0.78$; $p < 0.01$). Each dot refers to one population

Furthermore, there is no evidence of any regulatory mechanism (BAUER 1989). So the chance of a female leaving descendants increases with the number of glochidia produced. Especially since the survival rates of glochidia are so extremely low and density independent, a strong selection pressure must be expected to produce as many glochidia as possible. In general, a high number of progeny can be achieved in two ways (Fig. 2): either a high life expectancy leading to a high number of reproductive periods, or a high fertility leading to a high number of progeny per reproductive period. According to most concepts about life history selection these traits represent alternative strategies (GREENSLADE 1983, SCHAFFER 1974, SOUTHWOOD et al. 1974, STEARNS 1976), but the pearl mussel combined them. The combination of a high life expectancy up to more than 100 years (GRUNDELIUS 1987) and a fertility of some million glochidia per reproductive period (BAUER 1987b) yields a very high lifetime fertility.

However, this is already too general because the two important traits "life expectancy" and "fertility" are considerably plastic. Therefore, in the following chapter, the correlations between parameters related to lifespan and fertility are analyzed.

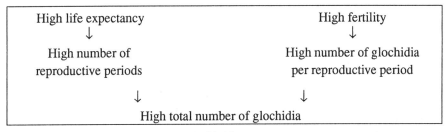

Fig. 2: Important traits in the pearl mussels life history strategy.

Plasticity

Growth studies

There is evidence that among poikilothermes life history traits like maximum size or life expectancy may be strongly influenced by the process of individual growth (BACHELET 1980, BEVERTON & HOLT 1959, GILBERT 1973, RAY 1960). Therefore growth studies were conducted along a latitudinal gradient covering nearly the whole north-south distribution of the pearl mussel. The analyzed populations are situated in Massachusetts (1 population), northern Spain (3), South Germany (16), Scotland (1) and Sweden (4). The geological substrate of all localities are primary rocks. In each population empty shells of all size classes were sampled, except for the Swedish ones, where the analyses are based on published data (GRUNDELIUS 1987, HENDELBERG 1961). The growth studies are based on the annuli. In order to make them visible, the shells were put in KOH (ca. 5%) at 50 °C, in this way removing the periostracum. With these shells three parameters related to growth were determined, namely the growth constant k, the maximum observed life span A_{max} and the maximum observed size L_{max}.

A basic parameter for the individual growth is the growth constant k, which determines the curvature of the average individual growth curve in a population, that is the rate at which the average asymptotic size is approached (BEVERTON & HOLT 1959). The lower its value the slower the growth. It was determined by measuring every fifth annulus along its longest axis. Then for each population, a regression was established relating shell length at age t+5 years to shell length at age t (Fig. 3, WALFORD 1946). The growth constant k is then:

$$k = -\ln (b), \quad \text{where } b = \text{slope of the regression}$$

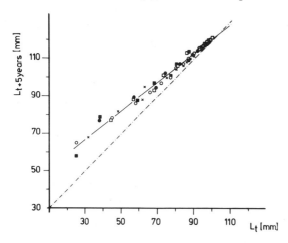

Fig. 3:   Example of a Ford-Walford Plot for one pearl mussel population. Each symbol refers to one analyzed shell. The equation of the regression is:
$y = 28 + 0.77x$; $r = 0.99$; $p < 0.001$.

Tab. 1: Correlation coefficients between growth constant, hydrochemical factors and latitude factors. (Between k and latitude the Spearmans rank correlation coefficient is given.

pH	BOD$_5$	Ca	Latitude
-0.65*	-0.78**	-0.67*	-0.78***

The maximum observed life span A_{max} is the maximum individual age recorded in a population. It was determined by counting the annuli. In this context it should be noted that under natural conditions mortality of adult pearl mussels is low, so that most mussels live out their alloted life span. Accordingly, the survivorship curve is convex (BAUER 1987 b). The maximum life span therefore must be considered as being more determined by intrinsic factors than by the rate of accidental mortality. Water pollution considerably alters the shape of the survivorship curve, but it has comparatively little effect on the maximum life span (BAUER 1988).

The third parameter, the maximum observed shell length L_{max}, indicates the maximum individual size attained in a population.

Fertility parameters

Fertility was estimated in nine populations. The number of glochidia per gravid female can easily be determined by keeping each mussel in a small bucket with only little water. This causes the release of glochidia which can then be counted. In contrast to most invertebrates there is usually no or at most a very weak relationship between fertility and age within a population (BAUER 1987b). Therefore, for each population an average fertility was calculated, i.e. the average number of glochidia produced by one gravid female per reproductive period. Since female pearl mussels do not reproduce every year (BAUER 1987b), the percentage of gravid mussels is also included in this analysis.

Correlations

The growth constant k shows relationships to a number of environmental factors (Tab. 1). It responds to hydrochemistry and to latitude. The latter relationship surely must be considered a temperature effect; at higher temperatures in the south, growth is accelerated.

When the relationships between k and the other traits are analyzed, the following pattern is evident (Fig. 4, Tab. 2):

At increasing growth constant the maximum life span decreases. Variation of this parameter is quite large; it ranges from 30 to 130 years (Fig. 4A). A similar relationship to k shows the maximum shell length (Fig. 4B), which again is related to fertility (Fig. 4C). In populations where the individuals are small, fertility is low compared to populations with

Tab. 2: Correlations between growth constant (k), maximum observed age (A_{max}), maximum shell length (L_{max}),fertility and percentage of gravid mussels.

	k	A_{max}	L_{max}	Fertility
A_{max}	-0.75***			
L_{max}	-0.55**	0.59**		
Fertility	-0.79*	0.55	0.84**	
% gravid	-0.40	0.11	0.16	0.37

large individuals. Table 2 shows, that the percentage of gravid mussels is not related to any of the other traits. On the average, every year 64 % of all females take part in reproduction and 36 % are put in a pause (BAUER 1987b).

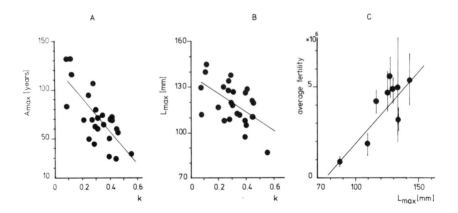

Fig. 4: Relationships between
A: growth constant k and maximum age A_{max} (y = 125 - 168x; r = -0.75; $p < 0.001$)
B: k and maximum shell length L_{max} (y = 139 - 62x; r = -0.55; $p < 0.01$)
C: L_{max} and fertility (= average number of glochidia per gravid female and reproductive period $\pm$ c.i.) (y = -6.2 + 0.08x; r = 0.84; $p < 0.01$).
Each dot refers to one population.

Plasticity and total reproductive output

Longevity and fertility are important factors for the number of glochidia produced by a female (Fig. 2). Since both parameters respond in the same way to k, A_{max} directly and fertility via L_{max} (Fig. 4, Tab. 2), a considerable reduction of the total reproductive output must be expected as k increases.

In Fig. 5 the total number of glochidia produced by one average female was calculated in relation to its growth constant k. The calculation is based on the relationships between k, A_{max}, L_{max} and fertility. Life expectancy is calculated according to a mortality rate of 10 % in ten years (BAUER 1988) and it is considered that the female pauses in 36 % of all reproductive periods. Furthermore, it is assumed that maturity starts when one fifth of the maximum life span has passed. This is largely an assumption. Since nearly all populations in Central Europe are overaged, this problem is very difficult to investigate. We know that in Scotland mussels attain an age of ca. 70 years and start reproducing at 12 - 13 years of age (YOUNG & WILLIAMS 1984). Presumably at increased growth constants the juvenile period is not reduced as much as is assumed here. But even in spite of this compensating mechanism a high growth constant leads to a large reduction of the total reproductive output. A female growing at k = 0.6 only attains roughly 10 % of the glochidia produced by a female which grows at k = 0.1.

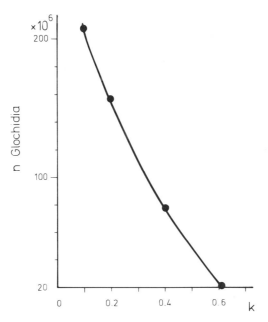

Fig. 5:   Total number of glochidia produced by one female pearl mussel in relation to its growth constant k

Plasticity and population growth

For the purpose of species conservation it is important to know how plasticity affects population growth. Therefore by means of a simple model the growth rate of the population (= r) is calculated in relation to the growth constant of the individuals (= k). The components of the model are presented in Fig. 6.

Adult stage (Fig. 6A)

The model operates with a constant age structure. As in the previous chapter, the mortality is 10% in ten years and maturity starts when one fifth of the maximum age has passed.

Glochidial and parasitic stage (Fig. 6B)

The complicated processes in this stage are combined in one equation. This equation describes the density of parasitizing glochidia out of the number of adult mussels. It is based on the data given in Fig. 1; however, a constant host fish density (the mean density recorded) is assumed in all rivers. The corresponding equation is y = -0,6 + 0,97x, (r = 0,79; p < 0.001). As the slope cannot be distinguished from b = 1, the model uses the equation y = -0.6 + x.

This equation can also be used to calculate the effect of different fertilities because two conditions are valid. The percentage of gravid animals as well as fertility are independent of age (BAUER 1987 b). The equation is based on field data from populations with an average fertility of 4 million glochidia per gravid female. If fertility x is used in the model, then log x/4 is added to the equation.

Furthermore, it is assumed that the river is 1 m wide, that the mussels are concentrated in one spot and that for 100 m downstream the chances of the glochidia finding a host are constant before dropping to zero. (The latter assumption surely is not far from reality, BAUER unpubl.).

Postparasitic stage (Fig. 6C)

The initial density of this stage is the density of parasitizing glochidia. (It has to be considered that the y-value of Fig. 7B only gives the density for one-year intervals.) A crucial point is the mortality during this stage as a mortality increase is the most important threat in Central Europe (BAUER 1988). YOUNG and WILLIAMS (1984) estimate this mortality in a **healthy** population at 95 %. To simulate the effect of an increased mortality the model uses rates of 95 %, 98 % and 99,995 %. The number of survivors finally yields the number in the first age class.

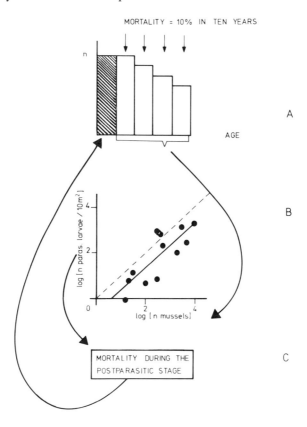

Fig. 6:   Schematic presentation of the model calculating the growth rate of a pearl mussel
population in relation to the growth constant of the individuals.
A = Adult stage;   B = Glochidial and parasitic stage;   C = Postparasitic stage
(See text for further explanation).

Calculation of r

For each k (0.1, 0,2 ... 0.6) A_{max} and fertility were calculated according to the
relationships in Fig. 4. With these values the growth rate of the population in ten-year
intervals was computed.

$$r = \ln \frac{N_{t+10}}{N_t}$$

The results (Fig. 7) suggest that there is a high risk of negative population growth if the
individual growth constant exceeds the value 0.5. Above this value there is a steep decrease
in growth rates. In this range there is also a strong effect of postparasitic mortality. Only at a
natural mortality of 95 % can positive growth rates be expected. Even a slight increase of
mortality leads to negative population growth. In general, the growth constant k seems to
have little effect at low values whereas at high values its effect on population growth is
tremendous.

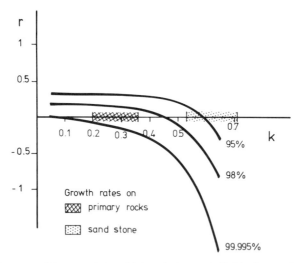

Fig. 7:   Growth rates of the population (r) in relation to individual growth constants (k) at three
           different mortality rates in the postparasitic stage.
           (The hatched and dotted area on the k-axis represent the 95 % c.i. of individual growth
           constants from two different localities in Germany).

Status of the freshwater pearl mussel in Central Europe

In many areas the freshwater pearl mussel formerly occurred at enormous densities
(Fig. 8).

Fig. 8:   Formerly the suitable rivers in the Fichtelgebirge were plastered with pearl mussels.
           (Photo by A. Ritter between 1930 and 1940).

In the Fichtelgebirge (North East Bavaria), suitable rivers were not only plastered for 20 - 30 km, frequently there were even two or three layers of mussels (ISRAEL 1913). However, since the beginning of the 20th century the numbers have decreased by more than 95 %. Nearly all of the remaining populations lack young mussels (Fig. 9) and will therefore become extinct within the next decades (BAUER 1983, WELLS *et al.* 1983).

The most important cause of decline is eutrophication, which already at very low levels increases the mortality of the postparasitic stage living in the river bottom. As Fig. 9 shows, in most cases the number of young mussels have decreased continuously during the last decades. Presumably the animals die because eutrophication leads to a high amount of detritus enriching the sediment with mud. In this way the river bottom becomes less and less suitable for the development of young mussels (BAUER 1988). The causal mechanisms leading to the death of the juveniles are unknown. Possible explanations are a shortage of oxygen or an increased predator density.

Also the adult stage is affected by eutrophication. Mortality at this stage is related to the nitrate concentration in the water (Fig. 10).

However, the adult stage is less sensitive than the postparasitic stage, i.e. as eutrophication increases first survival of the postparasitic stage is prohibited, then mortality of the adults rises. Thus, there are a number of populations which lack young mussels but mortality of adults still is low (BAUER 1988).

Fertility is remarkably independent of eutrophication and a subtle mechanism causing females to switch to hermaphrodites at low densities ensures a high fertility even in very sparse populations (BAUER 1987b).

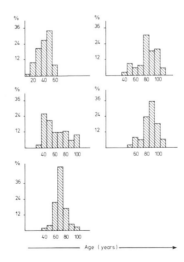

Fig. 9:   Some typical age structures of pearl mussel populations in Bavaria.

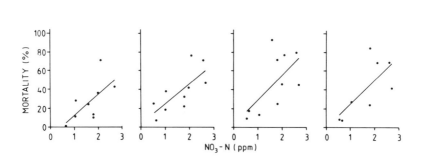

Fig. 10: Relationship between age-specific mortality of adult pearl mussels and the nitrate concentration in the water for eleven populations. (From BAUER 1988).

20 - 40 years :	y = -10.3 + 22.4 x;	r = 0,68; $p < 0.05$
40 - 60 years :	y = 3.3 + 21 x;	r = 0.74; $p < 0.01$
60 - 80 years :	y = 2.6 + 26 x;	r = 0.66; $p < 0.05$
80 - 100 years :	y = -0.85 + 26 x;	r = 0.74; $p < 0.05$

Influence of plasticity on status and conservation strategies

The results given in the last chapter undergo a considerable differentiation when plasticity is taken into account. In fact only one statement is independent of plasticity, namely that concerning fertility. Hence it follows for conservation strategies: With respect to fertility there is no critical density. Since there are probably no feedback mechanisms, the problem of a Minimum Viable Population does not seem to exist for the freshwater pearl mussel.

All further consequences only result from a combination of plasticity and threats. These consequences concern the danger of extinction (i.e. the sensitivity to threats), and conservation strategies.

The danger of extinction

Plasticity affects the danger of extinction via two components, namely

a) the growth rate of the population:

Eutrophication causing increased mortality, during the postparasitic stage should have a strong impact if the individual growth constant k exceeds 0.5 (Fig. 7). Already at a very low eutrophication level such populations should decrease and overage.

b) the time until extinction of overaged populations:

Most populations in Central Europe lack young mussels. The time until extinction of such populations is proportional to the number of age classes present in the population. Populations consisting of shortlived individuals (high values of k) will become extinct very soon. Of course the nitrate concentration in the water also has to be considered in this context.

There is an example for these hypotheses.

The results presented so far were gained from populations on primary rocks, which are the main distribution areas of the freshwater pearl mussel. Another suitable substrate are sandstones which are poor in lime. In Germany a distribution area on primary rocks is the Fichtelgebirge, whereas Odenwald, Spessart and Rhön are examples where the pearl mussel occurrs on sandstone (Fig. 11). Both areas are similar with respect to industrialization, human population density, etc. SEIDLER (1922) gives the maximum shell lengths for a number of sandstone populations. When they are compared to the maximum shell lengths of populations from the Fichtelgebirge, it is evident that the mussels on sandstone are smaller (Fig. 12).

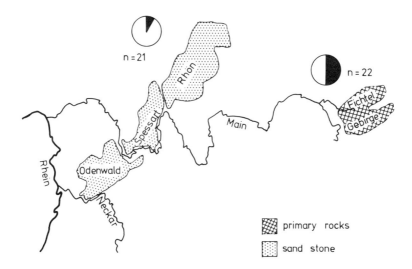

Fig. 11: Percentage of still populated pearl mussel rivers (black segments) in the Fichtelgebirge (primary rocks) and in sandstone areas (Odenwald, Spessart, Rhön). n = number of rivers which were originally populated. (Some rivers in the Odenwald were inhabited by mussels introduced from the Fichtelgebirge. They are not considered here).

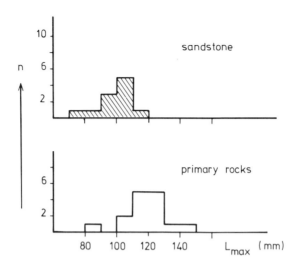

Fig. 12: Maximum shell lengths of pearl mussels on sandstone and on primary rocks (n = number of populations). The sandstone mussels are significantly smaller (t = 5.7; $p < 0{,}001$).

Converting their shell lengths into growth constants by means of the equation in Fig. 4B, yields an average of k = 0.62 ± 0.09, whereas in the Fichtelgebirge the average growth constant is 0.28 ± 0.09.

According to Fig. 7 a high danger of extinction must therefore be expected for the sandstone populations, whereas the populations in the Fichtelgebirge should be less vulnerable. This is confirmed by the present status (Fig. 11). Though in both areas nearly the same number of rivers had been populated, the pearl mussels have nearly disappeared from the sandstone mountains (HEUSS 1962, JUNGBLUTH 1986), whereas in the Fichtelgebirge half the populations still exist. Apparently, because of high growth constants associated with short life span and low fertility,the sandstone populations existed such that when conditions were optimal, the life history strategy barely ensured survival of populations. However, slightly worse conditions could not be tolerated and led to quick extinction. (Since most sandstone populations are extinct, I could investigate only one. The results were as follows: k = 0.63, A_{max} = 30 years, L_{max} = 88 mm and average fertility = 930 000 ± 30 000 glochidia per gravid female).

Conservation strategies

First of all, it is important that fertility depends neither on the age of the female, nor on population size. That means that all populations, even if they are very overaged or very small, are able to reproduce. There is no population which could not recover if effective conservation measures were taken.

Analysis of mortality showed that there are two weak points in the life cycle. One is natural, namely the extremely low and density-independent survival rates of glochidia. The other one is strongly influenced by man. It concerns the increase of postparasitic mortality by eutrophication. Accordingly, two conservation strategies may be proposed: Increase of the density of parasitizing glochidia by introducing infected fish and (or) improvement of the water quality. The first measure is simple and cheap. Up to 1000 Glochidia are able to develop on one trout fingerling. Therefore, it is possible to increase the number of young mussels released by the hosts considerably. Improving the water quality is of course always a desirable measure. However, it is expensive and at present can only be enforced in a few cases (BAUER & EICKE 1986). Since most of the remaining populations will become extinct within the next decades, conservation measures are urgently necessary. As there is no way to improve the water quality of all remaining pearl mussel rivers quickly enough, even delaying extinction would be a success in many cases. Thus, one has to decide which measure has to be taken in which case. Where is an artificial infection of host fish appropriate? Where is the only useful measure an improvement of the water quality? Which populations have high priority?

To solve these problems, the following procedure is suggested. (All measures refer to overaged populations). For each population the following parameters should be recorded:

Maximum shell length, maximum age, age structure, population size and hydrochemistry (especially conductivity, pH, nitrate, phosphate, calcium, and BOD_5 (BAUER 1988)).

a) Populations with large shells and long individual life span

These populations will still exist for many decades unless they are very overaged or the nitrate values are high. If the latter is true one must seek counter-measures.

In a large population (more than 5000 individuals), the natural infection intensities of the hosts and thus the initial densities of the postparasitic stage are high (Fig. 1). Every year many young mussels enter the river bottom, however none of them survive. This indicates that mortality in the postparasitic stage is absolutely 100 %. The river bottom is totally unsuited for young mussels. Thus, the only efficient conservation measure is an improvement of the water quality.

In a small population, the density of the postparasitic stage is also small. Even if there are some suitable spots in the river bottom, the chances are very small that one of these few young mussels will be released from the host just above one of them. Particularly if the water quality is not too bad, hosts which are artificially infected with glochidia from this population should be introduced.

b) Populations with small shells and short individual life span

These populations will soon become extinct; therefore they have the highest priority. Conservation measures must be taken very soon.

Already very low eutrophication levels must be expected to have a strong negative impact. In this respect a crucial weak point is the low fertility, leading to a low density of the postparasitic stage. As in the last case there is hardly any chance of one of these few young mussels surviving, even if the river bottom is not completely unsuitable. Therefore artificially infected fish should be introduced, no matter whether the population is large or small. If this measure is conducted for some years with many fish, (at least 1000 per population and year), it might be sufficient to provide a stock of young mussels able to survive the postparasitic stage. In this way the quick collapse of the population could be prohibited.

Conclusions

Detailed knowledge about the population ecology of a threatened species is an important requirement for effective conservation measures. However, as the example of the freshwater pearl mussel shows, it is problematic to apply demographic results gained from a few localities to other populations of the same species. The long lifespan of the pearl mussel, which is often mentioned in the literature (COMFORT 1957, HUTCHINSON 1979), does not hold for all populations. In addition reproductive parameters and sensitivity to threats are different. Only if the relationships between the plastic reactions of life history traits are known, can especially vulnerable populations become evident and a catalogue of conservation measures with broad validity be developed.

Zusammenfassung

Am Beispiel der in Mitteleuropa vom Aussterben bedrohten Flußperlmuschel (*Margaritifera margaritifera L.*) wird gezeigt, wie die Plastizität bionomischer Komponenten den Gefährdungsgrad einer Tierart beeinflußt. Die bionomische Strategie der Flußperlmuschel zielt auf eine möglichst hohe Glochidienproduktion ab, was durch die Kombination einer hohen Lebenserwartung (= hohe Zahl von Fortpflanzungsperioden) mit einer ebenfalls hohen Fertilität erreicht wird. Gerade diese beiden Komponenten zeigen aber starke plastische Reaktionen in Abhängigkeit von der individuellen Wachstumskonstante. Die Wachstumskonstante, d.h. die Geschwindigkeit, mit der sich die individuelle Wachstumskurve ihrer Asymptote nähert, wird von verschiedenen Umweltfaktoren beeinflußt: sie steigt z.B. mit zunehmender Temperatur. Eine Erhöhung der Wachstumskonstante (= "schnelleres Wachstum") führt zu einer Verringerung von Lebenserwartung, Maximalgröße und Fertilität. Dieser Mechanismus hat eine starke

Reduktion der gesamten (=lebenslangen) Glochidienproduktion bei schnellwüchsigen Tieren zur Folge. Für Populationen aus schnellwüchsigen Individuen muß daher eine hohe Empfindlichkeit gegenüber Gefährdungsfaktoren erwartet werden, wobei zwei Gründe besonders wichtig sind: ein verringertes Populationswachstum und eine aufgrund der geringeren Lebenserwartung verkürzte Zeit bis zum endgültigen Aussterben. Diese Hypothese wird durch die Bestandsentwicklung in verschiedenen Gebieten der BRD gestützt. Die Ergebnisse werden im Hinblick auf Schutzstrategien diskutiert

Acknowledgements

I am grateful to the "Fachberatung für das Fischereiwesen" in Oberfranken and Niederbayern for electrofishing, to D. SMITH and to F.R.Woodward for sending me shells and to Mrs. C. Vogel-Bauer and Mrs. S. Hochwald for comments . Mrs.A.Servant-Miosga provided much practical help. The manuscript was kindly typed by Mrs. G. Lutschinger and Mrs. M. Preiß.

References

BACHELET G. (1980):Growth and recruitment of the Tellinid Bivalve *Macoma balticaat* the southern limit of its geographical distribution. Mar. Biol. 59:105 - 117.

BAUER G. (1983):Age structure, age specific mortality rates and population trend of the freshwater pearl mussel. Arch. Hydrobiol. 98:523-532

BAUER G. (1987a):The parasitic stage of the freshwater pearl mussel, III. Host relationships. Arch. Hydrobiol. Suppl. 76:413-423

BAUER G. (1987b):Reproductive strategy of the freshwater pearl mussel. J. Anim. Ecol. 56:691-704

BAUER G. (1988):Threats to the freshwater pearl mussel in Central Europe. Biol. Cons. 45:239-253

BAUER G. (1989):Die bionomische Strategie der Flußperlmuschel. BIUZ 19,3:69-75

BAUER G., EICKE L. (1986):Pilotprojekt zur Rettung der Flußperlmuschel. Natur u. Landschaft 4:140-143

BERVEN K. A., GILL D. E. (1983):Interpreting geographic variation in life-History traits. Am. Zool. 23:85-97

BEVERTON R. J. H., HOLT S. J. (1959):A review of the lifespans and mortality rates of fish in nature, and their relation to growth and other physiological characteristics. CIBA Foundation Colloquia on Ageing 5: 142-180

BURGMANN M. A., AKCAKAYA H. R., LOEW S. S. (1988):The use of extinction models for species conservation. Biol. Cons. 43:9-25

COMFORT A. (1957):The duration of life in molluscs. Proc. Malac. Soc. Lond. 32:219-241

GILBERT M. A. (1973):Growth rate, longevity and maximum size of Macoma baltica. Biol. Bull. 145:119-126

GREENSLADE P.J.M. (1983):Adversity selection and the habitat templet. Am. Nat. 122:352-365

GRUNDELIUS E. (1987):Flodpärlmusslans tillbakagang i Dalarna. Information fran sötvattans - laboratoriet Drottningholm 4

HENDELBERG J. (1961):The freshwater pearl mussel. Rep. Inst. Freshw. Res. Drottningholm 41:149-171

HEUSS K. (1962):Die Flußperlmuschel. Natur und Museum 92:372-376

HUTCHINSON G. E. (1979):An introduction to population Ecology. Yale University Press, London

ISRAEL W. (1913):Biologie der europäischen Süßwassermuscheln. KG Lutz Verlag. Stuttgart

JUNGBLUTH H. J. (1986):in Bischoff W D, Dettmer R, Wächtler K (1986) Die Flußperlmuschel. Naturhist. Mus., Stuttgart.

RAY C. (1960):The application of Bergmann's and Allen's rules to the Poikilotherms. J. Morph. 106:85-108

SCHAFFER M. L. (1981):Minimum population sizes for species conservation. Bio. Science 31:131-4

SCHAFFER W. M. (1974):Optimal reproductive effort in fluctuating environments. Am. Nat. 108:783-790

SEIDLER A. (1922):Die Verbreitung der echten Flußperlmuschel im fränkischen und hessischen Buntsandsteingebiet. Ber. wett. Ges. Naturk.:83-125

SMITH-GILL S. J. (1983):Developmental plasticity: Developmental conversion versus phenotypic modulation. Am. Zool. 23:47-55

SOLOMON M. E. (1968):Logarithmic regression as a measure of population density response. Ecology 49:357-358

SOUTHWOOD T .R .E., MAY R. M., HASSEL M. R., CONWAY G. R. (1974):Ecological strategies and population parameters. Am. Nat. 108:791-804

STEARNS S. C. (1976):Life history tactics: a review of the ideas. An. Rev. Ecol. Syst. 8:145-171

WALFORD L. A. (1946):A new graphical method for describing the growth of animals. Biol. Bull. Mar. Biol. Lab. Woods Hole 90:141-147

WELLS S. P., PYLE R., COLLINS N. (1983):The IUCN Red Data Book. Gresham Press, Old Woking

YOUNG M., WILLIAMS J. (1984):The reproductive biology of the freshwater pearl mussel in Scotland I. Field studies. Arch. Hydrobiol. 99,4:405-422

Allozyme Variation and Conservation: Applications to Spiny Lobsters and Crocodilians

R. A. Menzies,   Institute of Evolution, University of Haifa, Haifa, 31999, Israel;   and:
* Dept. of Psychiatry, Div. of Psychoimmunology, University of South
Florida, Tampa, Florida 33613, USA

Abstract

Man's ability to promote extinction, either directly or through environmental stress or pressure, is well documented. His successes have also been recorded in the demise of "renewable resource" industries such as fisheries. Workable approaches to reversing these trends for a given species depend upon knowledge of a variety of natural history parameters such as population structure and dynamics. Identification of the geographic domains of breeding units or stocks is essential for the definition of management units. Application of allozyme studies has been successful in providing this useful information. The existence of significant gene frequency differences between populations is consistent with genetic isolation.

The spiny lobster, *Panulirus argus*, although originally presumed to be panmictic through most of its range, now appears to have population substructure with respect to gene flow. Observed genetic differentiation is consistent with a hypothesis of local selection pressures. Lack of absolute isolation of populations argues against genetic drift as a significant differentiation mechanism. These data suggest conservation strategies should be considered on a local or regional basis, as well as throughout the range.

Related studies on American crocodilians also reveal a degree of genetic differentiation between populations. Among the populations of *Alligator mississippiensis* studied there is little or no gene flow. The same is true for *Crocodylus acutus*. The level of genetic differentiation observed in the latter case may be a result of selectional response to local environmental stresses. Genetic variability, e.g., degrees of heterozygosity, was high (up to 15%) in the three populations sampled and most likely would not be a negative factor in long term adaptability. However, if genetic differences are responses to the environment, restocking strategies in conservation and recovery of depleted populations of this species may not be advisable. Other methods are described. On the other hand, because of the very low genetic variability in the alligator, both within and among populations, this should not be a problem should this species become endangered again. In fact, it may be advisable on a limited experimental basis to introduce individuals with genetic differences from various populations to a single experimental site.

---

* Send Correspondence to this address

Introduction

The ability of man to promote species extinction or processes leading to extinction is well documented. Recorded are numerous partial and complete successes such as various species of birds, sea turtles, cetaceans and others. His success has also been recorded in the demise of "renewable resource" industries, e.g., fisheries, such as the Caribbean conch, *Strombus gigus* and related species, various species of Palinurid lobsters, California abalone, etc., to mention a few.

The difference between over exploitation of many species such as whales and lobsters, is one of degrees based largely on the length of the reproductive cycle, fecundity and differences in the efficiency of harvesting of biomass. Impact has been either directly on adults or on early life cycle stages or both. Some species such as crocodiles and marine turtles have been hunted almost to extinction as adults and juveniles. With turtles, the impact is increased through the wholesale collection of eggs and destruction of nesting habitat. In Florida as elsewhere, juvenile and adult habitat of the spiny lobster has been lost to human land use and pollution. The same is true for crocodiles. Alligators have been more resilient by having success at cohabiting with humans, e.g., canals and lakes in developed areas. However their range has been fragmented by a variety of obstructions such as roads and drainage of wetlands.

Successful approaches to reversing trends that lead to a species or population extinction depend upon knowledge of numerous natural history parameters: critical stages of life cycles, essential ecological features or environmental requirements, population structure, dynamics, and especially recruitment. Identification of the geographic domains of breeding units or stocks and the sources of their recruits is essential. In other words, in order for conservation strategy to be successful, management units must be defined in terms of both stocks and geographic domains of interaction.

The focus of this paper is the application of allozyme techniques to assist in the identification of possible management units. The principle upon which this approach is dependent is the stability of gene frequencies in time, in animal populations with large effective size. Thus, the observation of genetic differentiation between two populations indicates restricted gene flow and, therefore, effective recruitment isolation. The two populations in this case could be considered as different management units.

Many investigators have suggested that adaptiveness is positively correlated with genetic variability e.g. heterozygosity (VALENTINE, 1976; VAN VALEN, 1965; LAVIE & NEVO, 1983 a,b.; NEVO et al., 1983; NEVO et al., 1986, others). The application of allozyme techniques also allows an assessment of genetic variability and, thus, may also reflect the genetic potential to avoid extinction.

Three studies are reported here in which genetic differentiation and variability have been assessed. Discussed will be the Caribbean spiny lobster, *Panulirus argus*, American alligator, *Alligator mississippiensis*, and the American crocodile, *Crocodylus acutus*.

The spiny lobster, *Panulirus argus*, was thought to be panmictic through most of its range because of the long planktonic larval period and the direction and velocity of oceanic currents. The presumption was the Caribbean basin and Florida Straits would create a gene mixing pool. Presented here are data not consistent with that scenario. Allozyme studies indicate that this species has a population substructure which could serve as the basis of defining regional or local management units.

Related studies on American crocodilians (*Alligator mississippiensis* and *Crocodylus acutus*) also reveal genetic differentiation between populations. However, degrees of heterozygosity are quite different for each species, suggesting the possibility of different conservation strategies.

Experimental Approach

Specimen collection and preservation: The details of procedures have been reported elsewhere (MENZIES & KERRIGAN, 1979; MENZIES, 1981; GARTSIDE *et al.*, 1977; MENZIES *et al.*, 1979; MENZIES & KUSHLAN, 1990; DESSAUER & MENZIES, 1984; DESSAUER *et al.*, 1984). Lobsters were collected in the field, frozen intact and transported to the lab for dissection and further processing. Blood samples from either alligators or crocodiles (GORZULA *et al.*, 1976) were obtained in the field and centrifuged as soon as possible to separate blood cells from plasma. When possible, a muscle biopsy was obtained from the thigh of alligators or crocodiles by the same procedure used for sea turtles (MENZIES *et al.*, 1983). Specimens were usually transported to the lab frozen. Figure 1 shows the location of the principle populations discussed in this paper. Alligator populations are labelled A; Crocodile, C; and Lobster, L.

Allozyme techniques:

Either vertical starch or polyacrylamide gel electrophoresis was performed on tissue extracts, blood cells or plasma. Enzyme localization procedures followed were as described by HARRIS & HOPKINSON (1976). All gel runs were done with internal standards, e.g., specimens whose banding pattern and mobility were known. Unknowns were scored blind. Specimens showing spurious staining indicative of post mortem degradation were discarded. The number of putative loci studied in our laboratory varied with each species; alligator, 44; crocodile, 32 and 18 for the lobster. In most cases, however, only a few loci, that is, polymorphic loci, were relevant to the calculations concerning genetic differentiation of populations.

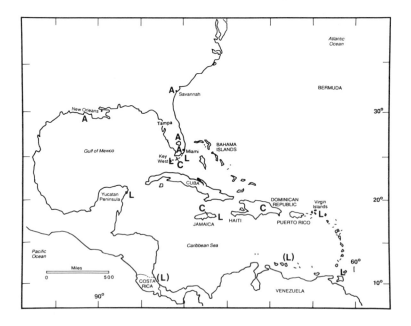

Fig. 1:   Map of the Caribbean Basin and Southeastern United States indicating the location of
various animal collections; A, alligator; C, crocodile and L, for lobster.

Calculations:

Assessment of genetic variability was done by averaging the number of heterozygotes over all loci, e.g., observed degree of heterozygosity (H), and calculation of the variance by standard methods (SOKAL & ROHLF, 1969). Estimated average degree of heterozygosity per locus, H, was also calculated according to NEI (1975, 1987). Polymorphic loci were tested for their fit to Hardy-Weinberg expectations by Chi-square tests. Where appropriate, e.g. small sample sizes, expected frequencies for each genotype were calculated by the method of LEVENE (1949). Genetic differentiation between populations was assessed in several ways. Genetic distance (D) as described by NEI (1975, 1987) but modified by HILLIS (1984) and TOMIUK & GRAUR (1988), D*, was calculated. Tests of significance were done using the variance calculated as TOMIUK & GRAUR (1988) described. Comparison of allele frequencies between populations were done calculating the variance as described by KEMPTHORNE (1957); $var(fx_i) = (fx_i)(1-fx_i)/2n$. Tests of significance were done by the t-test. Comparisons of putative genotype frequencies were done by either Chi-square or the G statistic (SOKAL & ROHLF, 1969).

Tab. 1:   Genetic variability in various alligator populations

POPULATON	LOCI	P	H	V(H)
LOUISIANA[1]	49	0.061	0.021	0.00014
LOUISIANA[3]	27	0.074	0.012	0.00002
FLORIDA[2]	44	0.023	0.009	0.00007
FLORIDA[3]	21	0.191	0.022	0.00004
SOUTH CAROLINA[3]	27	0.186	0.022	0.00004

P   = Proportion of loci whose most comon allele has a frequency less than or equal to 0.99.
H   = Heterozygosity
H(V)= Sampling variance of H
[1]   GARTSIDE et al. (1977).
[2]   MENZIES et al. (1979).
[3]   ADAMS et al. (1980).

Results and Discussion:

Alligator studies:

By 1977 the American alligator had been protected for a number of years. In many areas it had made a dramatic comeback. So successful was its recovery in the Rockefeller Refuge in Louisiana that a hunt was allowed to assist in managing the population. This enabled the group at Louisiana State University to do the first study of the genetic structure of a crocodilian population (GARTSIDE et al., 1977). The results were remarkable in that the degrees of heterozygosity was found to be a surprisingly low 2.1%, even though 49 putative protein loci were examined in 80 animals. Most verterbrates of this size have degrees of heterozygosity in the range of 5 to 10% (NEVO, 1978). It was not clear if the low degree of heterozygosity was a result of a founder effect after a bottleneck in a severely depleted population or the natural state of millions of years of alligator evolution perhaps greatly involving directed selection. To examine these possibilities we did a similar study on a collection of 63 alligators from a single location in Everglades National Park, more than 1500 km from the Rockefeller Refuge. Animal sizes ranged from 3 to 9 feet indicating a number of year classes. This coupled with knowledge of alligator mortality and movements in Everglades National Park and adjacent areas was consistent with the collection not being from a small number of clutches or parents.

Our findings were essentially the same as those of GARTSIDE et al., (1977) in that we recorded a degree of heterozygosity of 0.86% after examining 44 loci (MENZIES et al., 1979). In a later study ADAMS et al. (1980) observed the same pattern in three populations including the Rockefeller Refuge population. In all, the degrees of heterozygosity were not significantly different from each other and ranged from 0.86 to 2.2% (see Table 1). Since

Tab. 2:   Polymorphic loci from Louisiana (LA.)[1] and Florida (FLA.)[2] Alligator populations

| PROTEIN | GENOTYPES | | | ALLELE FREQUENCIES | |
POPULATION	a/a	a/b	b/b	a	b
LDH-2					
LA.	55	20	5	0.81	0.19
FLA.	35	26	2	0.76	0.24
CATALASE					
LA.	16	35	25	0.42	0.58
FLA.	0	0	63		1
TRIPEP-1					
LA.	4	27	49	0.22	0.78
FLA.	0	0	63		1
PGM					
LA.	0	0	80		1
FLA.	0	1	62	0.008	0.992

[1]   GARTSIDE et al. (1977).
[2]   MENZIES et al. (1979).

the Florida populations have never been as reduced as those of the Rockefeller Refuge or South Carolina, these data indicate that the low genetic variability is not due to recent bottlenecking events. Separate bottleneck events are also unlikely since all fixed loci in all populations are fixed for the same allele. These observations are consistent with directed selection over a very long period.

The polymorphisms presently existing in alligator populations might be a result of not only genetic drift but also local selection. At least one to several loci for each population pair shows genetic differentiation. Criteria for significance was either t-tests comparing allele frequencies or Chi-square tests comparing genotypic frequencies. In some cases a locus polymorphic in one population was apparently monomorphic in another (see Table 2).

If, as it has been suggested by others, low degree of heterozygosity or genetic variability increases a species vulnerability to extinction (VAN VALEN, 1965; VALENTINE, 1976), then the alligator is extremely vulnerable. On the other hand, the alligator might represent a species that has evolved a plasticity of physiological homeostatic systems allowing it to survive micro environmental fluctuations in an otherwise narrow but stable niche. This niche, geographically situated in the fresh water wetlands of the south east and south central regions of the north American continent, with the exception of human encroachment and intervention, has remained stable for millions of years.

The perception of low genetic variability is not entirely accurate. Ten loci were polymorphic in four populations studied. It is likely that if other populations and loci were studied, additional polymorphic loci would be detected. If there were greater communication, e.g., gene flow between populations presently blocked by human encroachment, overall genetic variability would be increased. As part of a long term conservation strategy, animals from various populations with different polymorphisms could be introduced to a controlled site. If average heterozygosity of that population increased, vulnerability to extinction might also decrease.

Crocodile studies:

Quite a different picture emerges from results of a genetic survey of several American crocodile populations. This work, which was a natural extension of the alligator studies, evolved from concern for the fate of the small, possibly dwindling, population of crocodiles in south Florida. Two questions were asked: Was genetic variability also low in the American crocodile as in the alligator? What might the impact be on the indigenous gene pool if a restocking strategy were implemented? Currently, there are only a few hundred crocodiles in the south Florida population. It would not take many immigrants to have an impact on the gene pool if significant genetic differences existed. If the differences were a response to local selection pressures, contamination with outside alleles could be disastrous.

Three populations were sampled: the Florida population, a captive Jamaican collection and a captive Dominican Republic collection. A total of 32 loci were studied. Although 25, 21 and 35 individuals were sampled from each population respectively, it was not always possible to resolve all loci in all individuals. Nevertheless, the results were very surprising in that degrees of heterozygosity for two of the populations, Florida and Dominican Republic, were in the range of 11.7% to 16% for the observed and estimated average degree of heterozygosity. The degree of heterozygosity for the Jamaican population was 5.5% for the observed and 8.1% for the estimated. These data are summarized in Table 3.

When tests for Hardy-Weinberg fit were done for each polymorphic locus on a population basis, 11 of 50 possible tests revealed significant deviations. Nine of these exhibited homozygote excess and two exhibited heterozygote excess. There was no obvious pattern distributing these loci either among populations or among loci. However, the deviations from expected may be an artifact of small sample size, although the appropriate correction (LEVENE, 1949) was made. When the test was done on pooled homozygotes and pooled heterozygotes for each locus, there were no significant differences.

Two criteria were used to judge genetic differentiation between populations. First, we calculated NEI's genetic distance (NEI, 1985, 1987) and variance, D*, V(D*), as modified by HILLIS (1984) and TOMIUK & GRAUR (1988). The calculations were done both by including all 32 loci (Table 4) and only polymorphic loci. Both methods yielded genetic distances significantly different from zero.

Tab. 3:   Genetic variability of three populations of *Crocodilus acutus*

POPULATIONS

VARIABILITY PARAMETERS	JAMAICA	FLORIDA	DOMINICAN REP.
% POLYMORPHISM[1]	28.1	50.0	50.0
OBS. AVG. HETEROZYGOSITY (PER LOCUS)[2]	0.055	0.117	0.158
VAR (OBS. AVG. HET.)[2]	0.009	0.017	0.039
"H" (ESTIMATED AVG. HET.)[3]	0.0813	0.1600	0.1492
VAR(H)[4]	0.0005	0.0009	0.0009

[1] Locus considered polymorphic if frequency of most common allele < 0.99.

[2] Mean and variance of the number of heterozygotes per locus.

[3] Estimated average heterozygosity per locus; $H = \sum_{I=1}^{r} h_I/r$;

   r loci; I^{th} locus

[4] Sampling Variance of H; $Var(H) = \sum_{I=1}^{r}(h_I - H)^2/(r(r - 1))$;

   $h_I = \sum (1 - x^2_{Ii})$ where x_{Ii} is the frequency of the ith allele of the Ith locus.

   For H and Var(H) see NEI (1975) p. 131.

The second calculation was a population pairwise comparison of the frequencies of the most common allele at a particular locus using a t-test, assuming unequal variance and n. Variances were calculated according to KEMPTHORNE (1957). Three loci were different between Florida and Dominican Republic, six loci between Jamaica and Florida, and nine loci between Jamaica and the Dominican Republic (Table 4). Genotype frequency differences between population pairs also revealed significant genetic differentiation.

The differences observed might be due to selection pressures indicating a possible local adaptational value. Another explanation in the case of the Florida and Jamaican collections is small effective population sizes. This may not be the case for the Lago Enriquillo population of the Dominican Republic which is quite large. On the other hand, if the gene pool there is dominated by only a few aggressive males, the effective population size could also be small.

From the degree of heterozygosity, it would seem that genetic variability is not a problem in terms of long term adaptability for any of the populations. However, in Florida

Tab. 4:   Genetic relatedness of three populations of *Crocodilus acutus*

Population Comparisons	D*[1]	V(D*[1])	LOCI WITH DIFF. ALLELE FREQ.[2]
Jamaica vs. Florida	0.0098	0.00001657	Est-2, PGM, CA, LDH-1, MDH-1, MDH-2
Jamaica vs. Dom. Rep.	0.0131	0.00002203	a2G, a3G, PER-2, NP, G6PD-2, PGM, CA, 6PGD, MDH-2
Florida vs. Dom. Rep.	0.0104	0.00001653	NP, G6PD-1, MDH-1

[1]   Genetic distance, D*, and variance, V(D*),
      calculated over 32 loci according to TOMIUK & GRAUR (1988).
[2]   Loci showing allele frequency differences based on t-test.

we appear to be dealing with a small relic population whose main survival problem is man's encroachment (KUSHLAN & MAZZOTTI, 1989). The existing range and habitat is nearly what it must have been before man or certainly this century. However, land use pressures in south Florida and increasing human contact, e.g. accidental and intentional killings, increase the threat to this small population. KUSHLAN (1988) has suggested a number of approaches. Very appealing is captive breeding and rearing. This has been successful for the Jamaican collection at Palmdale, Fla. (Cecil Clemons, proprietor). Restocking should not be considered as a first choice conservation strategy. Genetic differentiation between populations might be due to local selection. However, as an expediency, since no loci were fixed for a population specific allele, restocking should be considered if the Florida population is in danger of extinction.

Lobster studies:

The spiny lobster studies were conceptually and technically similar to the crocodilian studies in that certain basic population parameters needed definition in order to devise management plans. In this case, however, the focus was on prevention of extinction of an industry based on a renewable resource rather than on the extinction of the species. The specific objective was to define sources and rates of recruitment. It was hypothesized that an assessment of the genetic structure of populations might reveal patterns of gene flow and therefore recruitment patterns.

The spiny lobster, *Panulirus argus*, has a life cycle which includes a planktonic stage confounding the understanding of gene flow and recruitment. This is a common problem among many marine forms. Adult lobsters breed on reef tracts and release larvae (phyllosomes) to oceanic currents. The phyllosome larvae drift for 6 to 10 months developing through ten or eleven distinct morphological stages to eventually metamorphose to a puerulus or lobster looking animal better adapted for a benthic existence than the planktonic phyllosome. The puerulus, or post larva, settles in shallow near shore waters and takes 1.5 to 2 or more years to grow to sexual maturity to continue the cycle (Creaser, 1950; LEWIS, 1951; DAVIS, 1975; KANCIRUK & HERRNKIND, 1978; DAVIS & DODRELL, 1980; MENZIES & KERRIGAN, 1980).

The central question is, what is the source of post larvae that settle out in shallow waters? Considering the over all current flow through this specie's range, is there continual mixing and "down stream" transport (SIMS & INGLE, 1967; LYONS, 1981)? On the other hand, are there sufficient local gyres and counter currents present to enable a sufficient number of larvae to stay in the region of hatching? There do exist a number of counter currents and gyres throughout the Caribbean, Florida Straits and Gulf of Mexico that could account for at least regional retention of phyllosomes (MENZIES & KERRIGAN, 1979; MENZIES, 1981). Figure 2 illustrates the principle water transport patterns of the region. Not shown are a number of counter currents and gyres now known to exist. In the extreme, two models can be defined as either the "local recruitment" versus the "foreign recruitment" models (MENZIES & KERRIGAN, 1979). These two models have different genetic consequences. The former predicts a mixing of larvae from many populations creating a panmictic gene pool. Thus, down stream populations should not differ genetically from up stream populations. This is especially true if the system is not open ended, e.g., a loop back to the supposed origin. On the other hand, if recruitment is largely from local sources, e.g., local gyres and eddies returning a sufficient number of larvae, then there should be some distance of separation where populations are effectively isolated. Genetic divergence should

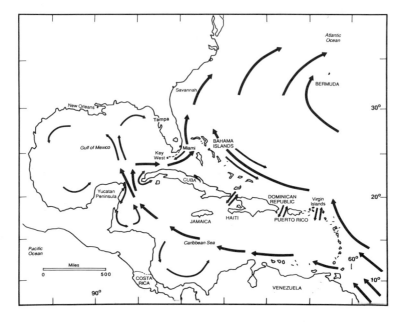

Fig. 2:   Map of the Caribbean Sea and Western Atlantic Ocean. Large arrows indicate major documented currents. Small arrows indicate minor currents and gyres.

occur by mechanisms such as selection or genetic drift. Thus, whether or not genetic differentiation exists will allow a test of these hypotheses.

Our initial approach was to collect adults and juveniles in a number of locations throughout the Caribbean and post larvae in several locations in Florida. We surveyed many protein loci in search of population specific fixed alleles which might serve as markers. We found none. However, we were able to identify a number of polymorphic loci that might reveal population specific gene or genotype frequency differences as was observed in both alligator and crocodile studies. The initial populations studied were from various Florida and Bahama locations, Belize and the Virgin Islands. At the outset we found numerous between population differences in genotype frequencies for several loci. The principle differences were between Belize and almost all other populations from Florida and the Bahamas. The Virgin Islands was distinct from several Florida but not Bahama populations. In this study the Bahama populations did not differ from each other, nor did the Florida populations. Further, with one exception, the Florida populations did not differ from those of the Bahamas. The only difference was between the two most northern studied, Boca Raton, Fla., and Walker's Cay, Bahamas.

These results indicated that genetic differentiation can be detected at the protein level between relatively distant populations but to a much lesser degree, if at all, between more closely situated populations. Thus, the dimensions in terms of both distance and numbers of polymorphic loci were in part defined for a successful assessment of pan-Caribbean relationships. Six strategically located populations and 8 polymorphic loci were chosen for broader study. The populations were chosen in part because of their position in the Caribbean and the direction of the main current flow. The populations were Elliot Key in the northern Florida Keys, Key West, Cancun on the Yucatan Peninsula, Jamaica, Virgin Islands, and Trinidad. The loci were chosen on the basis of preliminary data which suggested that their gene frequencies might show population differences. The loci were 3 peptidases, 2 esterases, a phosphatase, lactate dehydrogenase and phosphoglucomutase. The latter two were chosen as representatives of intermediary metabolism.

The assessment of genetic variability from these data is difficult since we introduced a bias by concentrating on polymorphic loci. The degree of heterozygosity calculated on 18 loci for which we had sufficient data, yielded values on the order of 14 to 16 percent. However, in our initial survey we studied over 20 additional loci which were either monomorphic in the few populations studied or very weakly polymorphic. Since these additional loci held little promise of contributing to the question of recruitment, they were not studied in all populations. If most of these were included in the genetic variability calculation, the expected degree of heterozygosity per locus would be about 7 to 8 percent. This is in the range expected for marine invertebrates (NEVO, 1978). For the purposes of assessing population differences, an accurate determination of degree of heterozygosity was not needed.

Tab. 5:   Genetic Distance Between Various *Panulirus argus* Populations.

Population Pairs	D*[1]	V[2]	poly D*[3]	V[4]
Elliot Key x Key West	0.0109	0.000043	0.0247**	0.000036
Elliot Key x Cancun	0.0246	0.000240	0.0563**	0.000212
Elliot Key x Trinidad	0.0117**	0.000033	0.0265**	0.000025
Elliot Key x Jamaica	0.0231	0.000144	0.0527**	0.000115
Elliot Key x Virgin Is.	0.0092	0.000023	0.0208**	0.000018
Key West x Cancun	0.0099**	0.000014	0.0225**	0.000008
Key West x Trinidad	0.0133**	0.000030	0.0302**	0.000019
Key West x Jamaica	0.0209**	0.000094	0.0477**	0.000068
Key West x Virgin Is.	0.0200**	0.000052	0.0455**	0.000026
Cancun x Trinidad	0.0277**	0.000158	0.0635**	0.000114
Cancun x Jamaica	0.0198	0.000150	0.0450**	0.000130
Cancun x Virgin Is.	0.0361	0.000336	0.0832**	0.000271
Trinidad x Jamaica	0.0253	0.000371	0.0577**	0.000350
Trinidad x Virgin Is.	0.0148**	0.000043	0.0336**	0.000030
Jamaica x Virgin Is.	0.0466	0.000773	0.1080**	0.000701

[1]   Genetic distances (D*) considering all 18 loci were calculated as originally defined by NEI (1975) and modified by HILLIS (1984) and TOMIUK & GRAUR (1988).
[2]   Variances were calculated as described by TOMIUK & GRAUR (1988) considering all 18 loci.
[3]   Genetic distances calculated on only 8 polymorphic loci.
[4]   Variance calculated on only 8 polymorphic loci.
**   Significantly different from zero, $p = 0.05$ or less.

Genetic differentiation between populations was calculated in two ways: genetic distances based on allele frequencies and direct comparison of genotype frequencies. Genetic distances as defined by NEI (1975, 1987) and modified by HILLIS (1984) and TOMIUK & GRAUR (1988) (D*) were calculated for each population pair. Variances were calculated as described by TOMIUK & GRAUR (1988). These calculations were done first including all 18 loci, and then considering only the eight polymorphic loci. In the former case about half of the pairwise comparisons were significantly different from zero. When only polymorphic loci were included, all comparisons were significantly different from zero (Table 5).

Genetic differentiation between population pairs can also be shown in genotype frequency comparisons. Since the genotype is the unit of evolution, its use to assess genetic differentiation is appropriate (Hedrick, 1971). Thus, pairwise tests of homogeneity were done for the six populations using the G statistic described by SOKAL & ROHLF (1969). These data are presented in Table 6. Most of the comparisons were significantly different. Exceptions include Elliot Key and Key West and Key West and Cancun. These were not surprising because of their close proximity. The lack of difference in the comparison of Florida populations with that from Trinidad may be an artifact of small sample size of the latter as well as the conservative nature of the G test. On the other hand, these populations

Tab. 6: Genotype frequency differences between Spiny Lobster population; "G"[1] Polymorphic
         Loci

POPULATION	ELLIOT KEY	KEY WEST	CANCUN	TRINIDAD	JAMAICA
KEY WEST	36.98				
CANCUN	49.46*	24.20			
TRINIDAD	30.47	28.18	54.08*		
JAMAICA	78.71*	66.03*	48.74*	36.64*	
VIRGIN ISLANDS	40.87*	63.42	79.22*	54.89*	119.77*

[1] G Statistic calculated as described by SOKAL & ROHLF (1969).
* Significantly different, $p < 0.05$.

did show differentiation by the genetic distance test (Table 5) which was based on allele frequencies.

Both the genetic distance test and comparison of genotype frequencies establish the existence of clear cut genetic differences between most of the lobster populations over the range studied. The major limitation appears related to sample size, in particular Trinidad and the number of loci studied. Greater confidence in the allozyme results is derived from the results of MCLEAN *et al.*, (1982) and KOMM *et al.*, (1982). These two groups, using mtDNA fingerprinting techniques, established genetic differences among several Florida populations.

It seems clear that gene flow, and therefore effective recruitment, must come from local or closely related populations. Geographic distance appears to be a major determinant of differentiation since there is a significant positive correlation of genetic distance parameters with distance in nautical miles (MENZIES, 1981). Clearly, distance as an isolating mechanism must be modified by local water transport patterns. Further, to what extent do genetic drift and selection play a role in the mechanism of genetic differentiation? An indirect test is possible from a consideration of genetic identities and maximum migration rates allowable to maintain present genetic differences between populations. According to NEI (1975), for two populations in genetic equilibrium the relationship between genetic identity, I, migration rates, m, between the two populations and mutation rate, v, is:

$$I^* = (m_1 + m_2)/(m_1 + m_2 + 2v)$$

Here, we have used the genetic identity, I^*, modified as previously described. I^* varies from .99 to less than .7 per locus. Mutation rate can be taken as 10^{-6}. If migration is only down current, $m_2 = 0$. Considering I^* at .99, an average migration rate per year in excess of 2 in 10,000 would cause an increase in genetic identity, eventually approaching unity. However, other studies have presented data suggesting that larvae arrive in Florida from non-Florida locations (LITTLE, 1977; LITTLE & MILANO, 1980; LYONS, 1981; MENZIES, 1981). An estimate of the percentage of total post larval arrivals per year may be as high as 25%. These larvae must yet survive approximately another three years before they can contribute to the gene pool. If it is assumed that the foreign arrivals experience no greater mortality than locally derived larvae, one can calculate the approximate number of years it would take to erase allele frequency differences. Even for as great a difference as 0.5, if 25% non-local recruitement is an annual event and there is no selection, in 5 years or less the frequencies would be indistinguishable. Therefore, most or all larvae surviving to reach sexual maturity and breeding, and thus being true recruits and not just arrivals, must originate from populations with similar gene frequencies. The alternative is that local selection mechanisms are sufficiently intense to effectively eliminate unfavorable genotypes. Although not rigorously tested across all loci, gene frequencies for some loci at several locations in Florida have been stable for up to 5 years.

A simple calculation can illustrate that very high mortality can be tolerated. In an unfished population females can produce over one million larvae each spawn and can spawn twice in a year. Thus, if one in a million return and survive and reproduce, a steady state can be achieved. In a heavily fished population, however, such as most of the Caribbean and Florida Straits, up to 90% of the adults are harvested. In this case survival of locally bred recruits must be of the order of 10 to 20 per female to maintain the stability of the local population. If not, recruits from distant populations and perhaps with lower fitness could successfully contaminate the indigenous population, possibly lowering overall fitness. This would not be a problem from the fishery point of view as long as there was an unlimited supply of recruits. However, if this scenario were repeated throughout the range of *P. argus*, the production of larvae would gradually diminish and the fishery would collapse. This must be further exacerbated if environmental changes occur in which new genotype combinations would have a lower fitness. To avoid this, local management strategies must include both limiting the total catch, as well as increasing protection of breeding females to maximize fecundity.

Summary and Recommendations

Allozyme population genetics has been applied to three species to generate data in support of conservation plans or strategies. For the crocodilians the aim was for the protection of an endangered or threatened species. For the lobsters protection of an endangered industry was the goal. Although the data generated had many similarities there were differences in application. The principle similarity was that all three species exhibited a population substructure.

Alligators have been historically successful in the south and southeast United States. Present policies for protection from hunting and in many areas for limiting habitat encroachment has allowed populations to stabilize and to even increase in many areas throughout the range. Existing genotypes have undoubtedly evolved a physiologic plasticity well suited for the present range of natural environmental changes. However, because of low genetic variability, long term management strategies for the species should include a component to increase gene flow at least on a limited or controlled basis. This might restore some genetic variability possibly lost because of population subdivision due to both natural and man induced habitat destruction and encroachment interrupting gene flow.

The Florida population of crocodiles, as other crocodile populations, has a great deal of genetic variability. Genetic differentiation between populations indicates that restocking should only be done if expedient measures are required and preferably only from genetically similar populations where no loci are fixed for different alleles. KUSHLAN (1988) has described management approaches which are more appealing, such as breeding and rearing programs. Finally, to be successful a program of habitat protection and restoration is imperative particularly for the Florida population.

Lobster allozyme data indicate that genetic variability is probably not a negative factor in long term species survival. Short of a massive catastrophe this species is not in danger of extinction. Conceivably, even if fished to the collapse of the industry, pockets would exist throughout the range producing enough larvae to avoid extinction. If postlarval and juvenile habitat have not been destroyed, recolonization would be possible.

To avoid a lobster industry collapse the best policy would be a pan Caribbean attempt to protect females through their productive years as well as restricting harvest. Presently, in Florida over 90% of the harvestable population is taken and mostly in the first few months of the season. Perhaps a shorter and possibly fragmented season holds the solution, as well as limited entry of fishermen and gear limitations.

For political reasons a pan Caribbean policy would be difficult to implement. However, allozyme data indicate that a local or regional policy could work. At present the data are imprecise enough to delineate management units by less than several hundred km. in diameter. Although Florida could be managed independently, the genetic data and oceanographic data suggest major cooperation with the Bahamas would increase the probability of success. Without more data it is difficult to assess the impact of Cuban or Yucatan populations on Florida populations. If more precise data from all of these populations were available, it might be possible to predict water transport patterns that could be tested by physical oceanographic techniques.

For the rest of the Caribbean, we see Trinidad, Virgin Islands, Jamaica and Yucatan (Cancun and Belize) as having little impact on each others recruitment. Preliminary (unpublished) data suggest populations in Caribbean Costa Rica and Bonaire (Netherland Antilles) may be independent from each other and other populations studied.

From the studies reported, it is clear that allozyme techniques provide the ecologist and conservationist with a powerful tool for assessing critical population management parameters. When taken with ecological, oceanographic or other techniques where appropriate, the development of conservation strategies can be done with greater predictability, confidence and success.

Zusammenfassung

Die Fähigkeiten der Menschheit, die Vernichtung von Arten direkt oder indirekt durch Umweltbelastungen zu beschleunigen, ist gut belegt. Ihre Erfolge wurden auch im Niedergang von Industriezweigen dokumentiert, die erneuerbare Ressourcen nutzen, wie z.B. die Fischereiwirtschaft. Machbare Ansätze, diesen Trend für eine bestimmte Art umzukehren, hängen von der Kenntnis einer Reihe von Lebensdaten ab, wie z.B. Populationsstruktur und -dynamik. Die Identifizierung der geographischen Ausdehnung von Fortpflanzungseinheiten ist wesentlich für die Definition von Managementeinheiten. Mit Hilfe von Allozymuntersuchungen konnten diese Informationen gewonnen werden. Das Vorhandensein signifikanter Unterschiede in den Genfrequenzen zwischen Populationen ist ein Hinweis auf genetische Isolation. Die Languste, Panulirus argus scheint bezüglich des Genflusses in Unterpopulationen geteilt zu sein, obwohl man ursprünglich annahm, daß sie über den größten Teil ihres Vorkommens panmiktisch sei. Die beobachtete genetische Differenzierung stimmt mit der Hypothese lokaler Selektionsdrucke überein. Der Mangel vollständiger Isolation der Populationen läßt es vorteilhaft erscheinen, daß genetische Drift ein wichtiger Mechanismus der Differenzierung ist. Diese Daten lassen Schutzstrategien als sinnvoll erscheinen, die auf lokaler oder regionaler Basis, aber auch über das ganze Verbreitungsgebiet wirken.

Ähnliche Untersuchungen an amerikanischen Krokodilen zeigen ebenfalls eine genetische Differenzierung zwischen Populationen. Zwischen Populationen von *Alligator mississippiensis*, der hier untersucht worden ist, besteht wenig oder kein Genfluß. Das gleiche gilt für *Crocodylus acutus*. Das Ausmaß an genetischer Differenzierung, die bei letzterem beobachtet wurde, kann das Ergebnis einer Reaktion auf lokale Umwelteinflüsse sein. Die genetische Variation, z. B. der Heterozygotiegrad, war in den drei untersuchten Populationen hoch (bis zu 15%) und ist vermutlich nicht nachteilig für eine langfristige Anpassungsfähigkeit. Wenn jedoch genetische Unterschiede Reaktionen auf die Umweltbedingungen sind, ist es nicht ratsam zum Schutz oder zur Stärkung von dezimierten Populationen Individuen einzubringen. Es werden andere Methoden beschrieben. Beim Alligator andererseits sollte dies wegen der geringen genetischen Variabilität innerhalb und zwischen Populationen kein Problem sein, wenn diese Art wieder gefährder sein sollte. In der Tat könnte es ratsam sein, in begrenztem Umfang versuchsweise genetisch unterschiedliche Individuen von verschiedenen Populationen in einem Versuchsareal auszusetzen.

Acknowledgements

I would like to thank the many individuals who assisted in these studies and donated their time and/or the use of their facilities from Nova University, Ft. Lauderdale, Fla., Everglades National Park, Fla., Florida Department of Natural Resources and numerous other organizations throughout the Caribbean. I would specifically like to thank, M. Kerrigan, E. Menzies, S. Raney, J. Kushlan, and G. Davis for their direct and long term involvement. I am sorry that it is not possible to list all the many others that assisted in various aspects of the projects. Financial support for these studies was provided in part from grants from the Academy of Marine Sciences, Inc., Miami; NOAA, Offices of Sea Grant (Florida) Grant No. 04-7-148-44046 and the National Science Foundation, Grant No. DAR8009353.

References

ADAMS, S.E., M. H. SMITH & R. BACCUS. (1980): Biochemical variation in the American alligator. Herpetologica. 36:289-296.

CREASER, R.P. (1950): Repetition of egg laying and number of eggs of the Bermuda spiny lobster. Proc. Gulf and Carib. Fish.Inst. 2:30-31.

DAVIS, G.E. (1975): Minimum size of mature spiny lobsters, *Panulirus argus* at Dry Tortugas, Florida. Trans. Am. Fish Soc. 104:675-676.

DAVIS, G.E. & J.W. DODRELL. (1980): Marine Parks and Sanctuary for spiny lobster fisheries managenment. Proc. Gulf and Caribbean Fisheries Inst. 32:194-207.

DESSAUER, H.C. & R.A. MENZIES. (1984): Stability of macromolecules during long term storage. In Collections of frozen tissues: Value, management, field and laboratory procedures and directory of existing collections. ed. H.C. Dessauer and M.S. Hafner. p18-22.

DESSAUER, H.C., R.A. MENZIES & D.E. FAIRBROTHERS. (1984): Proedures of collecting and preserving tissue for molecular studies. In Collections of frozen tissues: Value, management, field and laboratory procedures and directory of existing collections. ed. H.C. Dessauer and M.S. Hafner. p23-27.

GARTSIDE, D.F., H.C. DESSAUER & T. JOANEN. (1977): Genic homozygosity in an ancient reptile (*Alligator mississippiensis*). Biochemical Genetics. 15:655-663.

GORZULA, S., C.L. AROCHA & C. SALAZAR. (1976): A method for obtaining blood by vein puncture in large reptiles. Copea. 1976: 838-839.

HARRIS,H. & D.A. HOPKINSON. (1976): Handbook of Enzymes Electrophoresis in Human Genetics. North Holland Elsevier Pub. N.Y.

HEDRICK, P.W. (1971): A new approach to measuring genetic similarity. Evolution. 25:276-280.

HILLIS. D.M. (1984): Misuse and modification of NEI's genetic distance. Syst. Zool. 33:238-240.

KANCIRUK, P. & W. F. HERRNKIND (1978): Reproductive potential as a function of female size in *Panulirus argus*. In Florida Sea Grant Tech. Paper No. 4. R. WARNER, ed. Univ. of Florida, Gainesville.

KEMPTHORNE, O. (1957): An Introduction to Genetic Statistics. John Wiley & Sons. New York.

KOMM, B., A. MICHAELS, J. TOSKOS & J. LINTON. (1982): J. COMP. Biochem. Physiol. 78B:923-929.

KUSHLAN, J.A. (1988): Conservation and Management of the American Crocodile. Environmental Management. 12:777-790.

KUSHLAN, J.A. & F.J. MAZZOTTI. (1989): Population Biology of the American crocodile. J. of Herpetology. 23:7-2.

LAVIE, B. & E. NEVO. (1981): Genetic diversity in marine molluscs: A test of the Niche-Width variation hypothesis. Mar. Ecol. (Berlin) 2:335-342.

LEVENE, H. (1949): On a matching problem arising in genetics. Annals Math. Stat. 20:91-94.

LEWIS, J.B. (1951): The phyllosoma larvae of the spiny lobster, *Panulirus argus*. Bull. Mar. Sci. 189-103.

LITTLE, E.J.,Jr. (1977): Observations on recruitment of postlarval spiny lobster. *Panulirus argus*, to the south Florida coast. Florida Dept. Natural Resources. Mar. Pub. No. 29. 35pp.

LITTLE, E.J.Jr. & G.R. MILANO. (1980): Techniques to monitor recruitment of post larval spiny lobsters, *Panulirus argus* to the Florida Keys. Florida Dept. Natural Resources. Mar. Pub. No. 37. 16pp.

LYONS, W.G. (1981): Possible sources of Florida's spiny lobster population. Proc. Gulf and Caribbean Fisheries Inst.31:253-266.

MCLEAN, M., C.K. OKUBO & M.L. TRACEY. (1982): mtDNA heterogeneity in *Panulirus argus*. Experientia. 39:536-538.

MENZIES, R. A., & J. KUSHLAN. (1991): Genetic variation in populations of the American crocodile, *Crocodylus acutus*. J. Herpetology. (in press).

MENZIES, R.A., J. KUSHLAN & H.C. DESSAUER. (1979): Low degree of genetic variability in the American alligator (*Alligator mississippiensis*). Isozyme Bull. 12:55.

MENZIES, R.A. & J.M. KERRIGAN. (1979): Implications of spiny lobster recruitment patterns of the Caribbean--A biochemical genetic approach. Proc. Gulf and Caribbean Fisheries Inst. 31:164-178.

MENZIES, R.A. & J.M. KERRIGAN. (1980): The larval recruitment problem of the spiny lobster. Fisheries. 5:42-46.

MENZIES, R.A. (1981): Biochemical population genetics and the spiny lobster recruitment problem: An update. Proc. Gulf and Caribbean Fisheries Inst. 33:230-243.

MENZIES, R. A., L. KOCHINSKY & J.M. KERRIGAN. (1983): Techniques for muscle biopsy and blood samples from sea turtles. Proc. Western Atlantic Turtle Symposium. U.S. National Marine Fisheries Service. Miami.

NEI, M. (1975): Molecular Population Genetics and Evolution. North-Holland Publishing Co., Amsterdam. 288pp.

NEI, M. (1987): Molecular Evolutionary Genetics. Columbia University Press, New York. 512pp.

NEVO, E. (1978): Genetic variation in natural populations: Patterns and theory. Theoret. Pop. Biol. 13:121-177.

NEVO, E. (1983):. Adaptive significance of protein variation. in Protein Polymorphism: Adaptive and Taxonomic Significance. ed. G.S. OXFORD & D. ROLLINSON. Academic Press. London and New York. p239-282.

NEVO, E., (1983):. Population Genetics and ecology: The interface. In: Evolution from molecules to man. D. S. Bendall (ed.) Cambridge Univ. Press. Cambridge, England, pp. 287-321.

NEVO, E., B. LAVIE, & R. BEN-SHLOMO. (1983): Selection of allelic isozyme polymorphisms in marine organisms: Pattern, Theory and Application. Isozymes 10:69-92.

NEVO, E., R. NOY, B. LAVIE, A. BEILES & S. MUCHTAR. (1986): Genetic diversity and resistance to marine pollution. Biological Jour. Linnean Soc. 29:139-144.

SIMS, H.W. Jr. & R.M. INGLE. (1967): Caribbean Recruitment of Florida's spiny lobster population. Quarterly J. Fla. Acad. of Sciences. 29:207-242.

SOKAL. R.R & F.J. ROHLF. (1969): Biometry. W.H. Freeman and Co. San Francisco, 776pp.

TOMIUK, J. & D. GRAUR. (1988): NEI's modified genetic identity and distance measures and their sampling variances. Syst. Zool. 37:156-162.

VALENTINE, J.W. (1976): Genetic strategies of adaptation. In Molecular Evolution, F.J. AYALA, Ed. Sinauer Assoc.,Inc., Sunderland, Mass. pp78-94.

VAN VALEN, Z. (1965): Morphological variation and width of ecological niche. Am. Nat. 99:337-390.

Conservation of Amphibian Populations in Britain

R.S. Oldham and M.J.S. Swan,
Department of Applied Biology and Biotechnology, Leicester Polytechnic, Leicester LE7 9SU ,
Great Britain

Abstract

Species conservation involves an evaluation of distribution and status, knowledge of habitat requirements, the extent of dispersal and specification of practical measures to ensure survival, together with an appropriate legal framework and public attitude.

Evaluation of amphibian distribution and status in mainland Britain has involved a national survey, the methods, findings and limitations of which are reviewed. Knowledge of habitat requirements and dispersal patterns essential for conservation, may be gained from survey questionnaire returns and by field experiments at the population level. Our work on *Triturus cristatus* and *Bufo bufo* is used as an illustration. Practical measures to effect conservation may include the use of reserves, tunnel systems, barriers to movement, translocation of populations and habitat management. These options are discussed with reference to the same two species. The interaction and impact of legislation and public attitude, both as important to conservation as biological aspects, are discussed. Amphibian conservation in Britain still has much to achieve, but the progress is not entirely discouraging.

Introduction

Species conservation involves an evaluation of distribution and status, knowledge of habitat requirements, the extent of dispersal and specification of practical measures to ensure survival, together with an appropriate legal framework and public attitude. This paper reviews our work on these aspects in respect of the British amphibia.

The British Isles has an impoverished amphibian fauna, relative to the rest of Europe; there are six indigenous species, three anurans *Rana temporaria*, the common frog; *Bufo bufo*, the common toad; and *B. calamita*, the natterjack toad; and three urodeles *Triturus vulgaris*, the smooth newt; *T. helveticus*, the palmate newt; and *T. cristatus*, the crested newt. All of them receive some protection under the "Wildlife and Countryside Act" (H.M.S.O.; 1981), which applies to England, Wales and Scotland. The sale of *R. temporaria*, *B. bufo*, *T. vulgaris* and *T. helveticus* is prohibited. For the other two species,

B. calamita and *T. cristatus*, there is a greater level of protection; it is illegal to capture, injure or kill them, or to possess them, alive or dead. More significantly, for these two species, it is forbidden to damage, destroy, or obstruct access to "any place of shelter". Exemption from the regulations is possible under licence from the Nature Conservancy Council (N.C.C.). Licences may be issued to permit scientific and educational work and in some cases, for instance when there are especially good reasons for commercial land development, to permit the translocation of populations. In theory the legislation therefore offers almost complete protection from human interference to two of the species. In practice, for a widespread animal like the crested newt the legislation cannot be rigidly enforced. Most members of the public would be unaware that by digging their gardens, for instance, they may be "destroying a place of shelter" of a protected newt species. There is controversy concerning the efficacy of the Wildlife and Countryside Act as applied to amphibia (e.g. BEEBEE 1988, COOKE & OLDHAM 1988); nevertheless, there can be little doubt that the legislation has enhanced public awareness of the amphibia and obliged developers to seek advice from the N.C.C. before tampering with habitats.

The natterjack toad is much more restricted than the crested newt. It is known to breed in only about 30 localities (COOKE, pers. comm.) and is classed as "endangered". Any planned incursions into its habitat by land developers are vigorously opposed by the N.C.C. The more widespread crested newt is classed as "vulnerable" and the degree of practical protection it receives from the N.C.C. depends upon the estimated size of the breeding population. The number of animals observed in the water during a circuit of the breeding site at night (a "torch count") is used as an indicator of population size. Any site with a torch count of over 100 is considered by the N.C.C. for designation as a Site of Special Scientific Interest (S.S.S.I.; N.C.C. 1989). Such sites usually remain in the original ownership but there are restrictions on the way in which both the aquatic and terrestrial parts of the habitat may be managed. In some cases a site is purchased by the N.C.C. on behalf of the nation; it then receives full protection and is designated as a National Nature Reserve (N.N.R.). Every natterjack toad site is designated as either an S.S.S.I. or an N.N.R.; but less than 1% of crested newt sites are so protected (SWAN & OLDHAM 1989).

At Leicester Polytechnic, under contract to the Nature Conservancy Council (N.C.C.), we are evaluating British amphibian resources, partly by assembling existing records from diverse sources and partly by stimulating and co-ordinating new surveys. The contract also demands an investigation of the ecological requirements of amphibian populations so that theirconservation may be placed on a firm ecological basis. Additionally, we have been involved with the provision of practical advice to ameliorate the damage which various development programmes (roads, housing, mining etc.) may cause to amphibian populations.

National Amphibian Survey

Financial resources committed to amphibian survey work are limited. Our contract provides a full-time salary for only the survey co-ordinator (MJSS), supplemented by small sums to reimburse surveyors' travelling expenses. Most of the survey work therefore depends upon voluntary effort and publicity is needed to stimulate interest. In 1988, for instance, our efforts in this direction were made via a N.C.C. press release. In turn this resulted in a series of articles and radio and television interviews. The media involved are categorised in Figure 1, which also shows the proportions of inquiries we received as a result of publicity in each category. The productivity of the publicity can better be judged, however, by the number of people actually submitting survey returns and better still by the productivity of those people. These aspects are illustrated in Figure 1, which shows that the most efficient form of publicity was obtained from natural history magazines (19% of the inquiries, 13% of whom submitted records with a mean productivity of more than 10 records per respondent. In the three years 1986 to 1989, we received inquiries from 2,493 people, 588 of whom submitted 6,484 site records.

Involvement of the general public carries the enormous advantage of widespread national coverage (Fig. 2) and the incidental but important benefit of increased public awareness of the existence and the status of the animals. It carries the disadvantage that survey cover is uneven, both qualitatively and quantitatively; there is a tendency for a disproportionate number of records to come from suburban areas and surveyors vary greatly in their expertise. These problems are superimposed on to the basic difficulty of assessing population success by observation of animals at night during a short, weather dependent breeding season. The problem of uneven survey cover, discovered in early years of the survey, could be corrected to some extent by targeting specific areas of the country in the publicity effort in subsequent years. The other limitations, however, cannot be easily overcome in the absence of increased resources, but they must be borne in mind in interpreting the survey results.

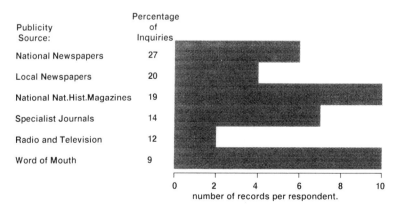

Fig. 1:   Variation in producivity of survey respondents to different publicity sources.

Information was requested at three levels, to accommodate differences of commitment in the surveyors: (1) **simple site records**, which simply list site locations and observed occurrence of amphibians; (2) **questionnaires**, which provide site descriptions as well as species and numbers observed; (3) **blanket surveys**, in which questionnaires are completed for all water bodies within a given area. The simple site records (49% of records) are of value in plotting species distribution, the individual questionnaires (20%) additionally provide a means to investigate habitat requirements, and blanket survey results (31%) permit the comparison of amphibian status in different parts of the country.

Amphibian Distribution

In this paper detailed results are not appropriate (these are provided by SWAN & OLDHAM 1989), but Figure 3 shows an example of a distribution map with the recorded occurrence of crested newts (solid circles) within 10km squares in the British Isles. The open circles indicate 10km squares in which ponds have been searched but no crested newts found. The species was found in 11% of sites examined and has a predominantly southern and eastern distribution, probably because in other areas, which are colder or have nutrient poor water, the aquatic sites are insufficiently productive to support populations. Figure 4

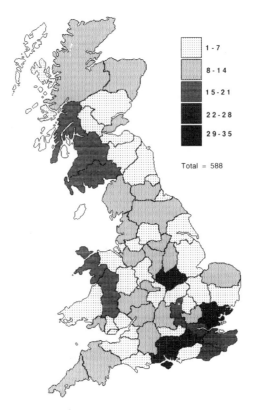

	1 - 7
	8 - 1 4
	1 5 - 2 1
	2 2 - 2 8
	2 9 - 3 5

Total = 588

Fig. 2:   Total number of amphibian recorders contributing to the NCC survey, 1986-9, per county.

shows, on a county basis, the relative proportions of ponds which were searched and yielded no evidence of amphibia. Counties with high values, for example in the eastern midlands and the north-west of England, are generally those in which there is intensive agriculture. The crested newt is vulnerable because the heart of its distribution occurs in areas where the aquatic habitats are under threat.

One of the questions asked of surveyors concerned the extent of damage and threat of damage to crested newt sites. The majority of sites actually destroyed were accounted for by land development projects (Fig. 5); urban development (52% of 23 sites between 1983 and 1986) and agricultural development (22%). In the same period, observers regarded 421 sites as being under threat. In this case, however, only 21% were affected by the two forms of land development combined (Fig. 5). By far the greatest perceived threat was pond senescence, the natural consequence of hydroseral succession. In earlier forms of agricultural practice succession was inhibited by regular clearance of silt, plant debris and excess macrophytes. Today ponds tend to be redundant agriculturally and are much more commonly neglected; succession carries the habitat beyond the stage that is optimal for newt productivity. For example 23% of ponds surveyed between 1986 and 1989 showed over 75% encroachment by emergent vegetation.

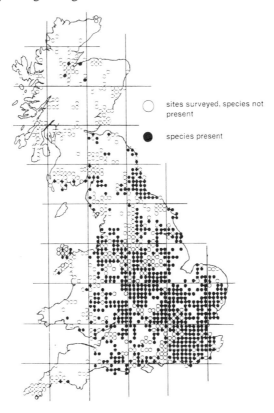

sites surveyed, species not present

species present

Fig. 3:   Distribution of *T. cristatus* throughout mainland Britain by 10 km grid squares.

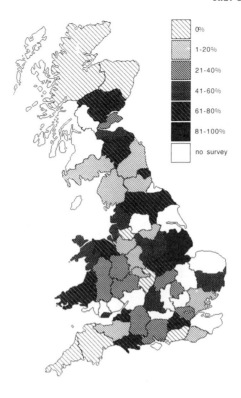

Fig. 4:   Percentage of sites recorded in "blanket" surveys in each county containing no amphibians at all.

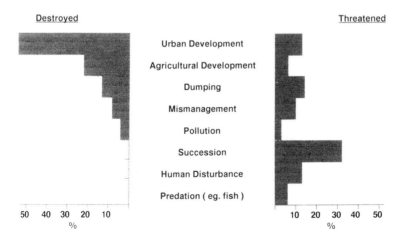

Fig. 5:   Factors cited as responsible for destruction of or threats to *T. crisatus* sites.

Habitat Requirements

Spatial Patterns Indicated by the National Survey

Inferences on habitat requirements can be gained from the questionnaire returns which document amphibian habitat usage (OLDHAM & NICHOLSON 1986, SWAN & OLDHAM 1989). For example, crested newts were found most commonly in the more mature farm ponds, with surface areas in the range 100 to 500 m^2, rich in nutrients and aquatic vegetation and characterised by occasional desiccation with associated freedom from fish. These sites were almost invariably shared with smooth newts, but less frequently with frogs and toads. Garden ponds are rarely appropriate, probably because of their small size and the frequent inclusion of fish. Populations were most commonly based in ponds surrounded by pasture land, but population studies indicated that this was an accident of pond distribution; ponds are frequently located in pasture as a source of drinking water for livestock, but the pasture itself is not used as a newt habitat. Scrub and gardens seemed to provide the most favourable terrestrial habitats, but good populations existed in a variety of habitat types, provided there was a mosaic distribution, containing an 'unmanaged' component, with dense ground cover.

Spatial Patterns Indicated by Field Experiments

Adult Influx

At best the information gleaned from questionnaires is circumstantial and it is necessary to undertake complementary field experiments to approach more closely to an understanding of habitat requirements. For example, since 1982 experiments have been conducted on a large toad population (7,000 adults, terrestrial density about 20 Ha^{-2}) in Coleorton, Leicestershire. The breeding site is a one hectare pond set in 7 Ha of woodland but surrounded by land managed in several markedly different ways (Fig. 6). To the north is thinly grazed scrub; to the west cultivated land, in some years ploughed, in other years grassed and intensively grazed; to the south-west, across a busy road, is moderately grazed pasture; to the south is woodland, pasture and scrub; to the south-east is a hay field; and to the east is more intensively grazed pasture. To the east and south-east, beyond the proximate fields, are gardens. Adult toads were intercepted during their influx to the pond in early spring, each year from 1982 to 1989, using drift fences (167 m in total) and pitfall traps around the distal boundary of the woodland. The fences also intercepted the adults' return movement in the late spring, as well as the exodus of the metamorphs during the summer.

Examples of the distribution of catches during the annual breeding influx are shown in Fig. 7. The northern scrub areas apparently supported the largest proportion of the population and this was a consistent feature from year to year. The western cultivated land supported a relatively small proportion of the population, again consistent from year to year despite dramatic changes in agricultural use (from ploughed arable land to intensively grazed pasture). This could be explained if the toads were using the hedges and ditches at

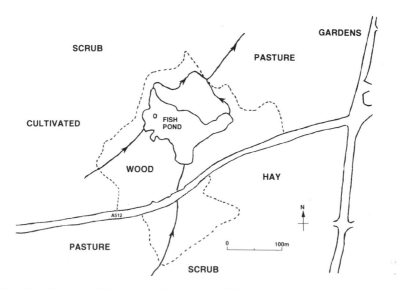

Fig. 6:   Toad breeding site, Coleorton, Leicestershire ("Fish Pond") with surrounding land use.

the margins of the fields; these constituted a small proportion of the total area, but provided a similar habitat each year. A segment of the habitat to the south-west was ungrazed scrub before autumn 1983, but was then ploughed, seeded and joined to the western pasture. The catches from this segment accordingly, declined dramatically. Animals from the south could gain access to the pond either across the busy A512 road or under a bridge. The proportion of captures arriving via the bridge showed a steady annual increase, whilst those arriving via the road declined. There were no obvious vegetation changes which might account for this pattern, but there was a steady increase in the volume of traffic on the A512 and it seems likely that an increase in toad road deaths caused the change. This is supported by the fact that an increased proportion of the road kills in more recent years has been located on the southern half of the carriageway indicating that few of the animals now cross the road successfully.

The eastern segment (Fig. 7), predominantly containing pasture, provided a variable proportion of catches, much higher in 1985 than in other years. Again, there is no clear habitat change that might explain the difference. Newly metamorphosed toads (metamorphs), captured in the traps as they dispersed during early summer also showed variation in numbers in the different habitat segments. In this case, however, the differences were related to conditions in the pond rather than to differences in the terrestrial habitats. The highest proportion dispersing towards the east was recorded in 1982 (33%). Subsequently the aquatic conditions in the north-eastern limb of the pond (Fig. 6) steadily deteriorated owing to a build up of organic detritus and the numbers of metamorphs leaving towards the east steadily declined. By 1986, for example, the numbers dispersing to the east had dropped to 15%. The proportions of adult toads entering the pond from the east was

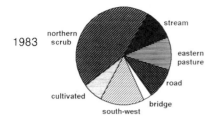

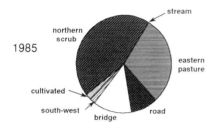

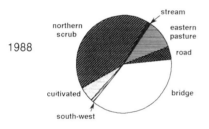

related to the proportions of metamorphs leaving towards the east three years beforehand. An understanding of terrestrial dispersion, therefore, needs to take into account aquatic conditions as well as terrestrial conditions.

Metamorph Dispersal

Metamorph dispersal from this site has been described by OLDHAM (1985). Emergence of metamorphs from the pond is completed within about three weeks and the pattern of dispersal is characterised by a steady outward progression, the animals moving more or less at right angles to the shoreline regardless of the habitat they encounter **en route** (Fig. 8). The fastest rate of travel occurred along roads (11m per day), the slowest through thick grass (1.5m per day). Outwards movement ceased with the onset of cold autumn weather.

Newly metamorphosed newts disperse widely during their first year, reaching distances of up to 500m (OLDHAM & NICHOLSON 1986). Return to the breeding site occurs at the end of either their first or second season and subsequent dispersal is probably less extensive; up to about 200m, but heavily dependent upon the habitat in the vicinity of the pond. If suitable habitat is remote from the pond there must be corridors connecting it to the pond. The crested newt

Fig. 7: Proportions of adult toads arriving at the "Fish Pond", Coleorton, from different locations, as indicated by catches at 167m of drift fencing.

spends more time in water than the other British species; nevertheless the importance of the provision of appropriate terrestrial habitat is emphasised by the fact that the immature newts spend at least one year entirely on land and even the adults in the 'aquatic' phase periodically return to the land.

Temporal Patterns of Movement

In the toad study the drift fences were set along the outer perimeter of the woodland around the pond (Fig. 6) so the catches reflect movements between the woodland and the surrounding habitats. An example of the annual pattern is shown in Fig. 9. The peaks in April represent the first seasonal mass movement, an influx of adult toads immediately prior to the breeding season. Breeding activity occupied about ten days, after which three weeks elapsed with very few captures. During this period the toads were probably on land but

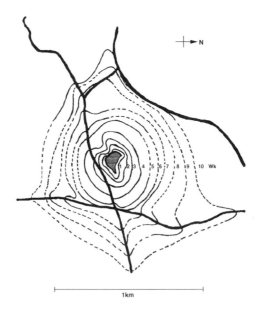

Fig. 8:   The progression of toad dispersal from the "Fish Pond", Coleorton as indicated by
           observation of metamorphs at weekly intervals in summer 1982. The contours represent
           the leading edge of the wave of dispersal (——— observation,   - - - -assumption).

remained inactive in the relatively low air temperatures pertaining at that time of the year, temperatures apparently adequate to induce breeding behaviour but not normal nocturnal activity. The second mass movement in the warmer weather of late May was the result of captures made during the exodus of adults from the vicinity of the pond. Another, higher peak of captures (the third mass movement), in early July, was the result of newly metamorphic toads moving away from the pond. Subsequent oscillations in capture numbers tended to correlate with rainfall pattern, animals moving into the woodland during dry periods and out in wet weather. From September 18th onwards, however, the slight increase in numbers captured (the fourth mass movement) was the result of a pre-winter influx of adults which seemed to anticipate the spring, positioning the animals close to the pond. Details of the temporal pattern varied from year to year but the four main phases of mass movement occurred consistently.

Practical Conservation Measures

Vulnerable Species

The protection of endangered species such as the natterjack toad involves the preservation of all available habitat. With vulnerable species which are much more widespread, such as the crested newt, there is rarely the opportunity to devote broad tracts of

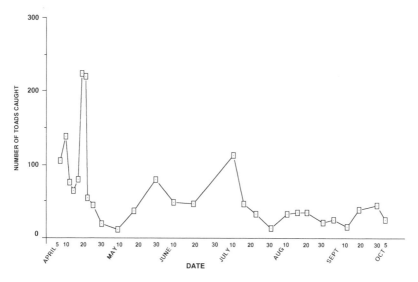

Fig. 9:   Number of toads caught in pitfall traps around the "Fish Pond", Coleorton, in 1982.

land exclusively to the species. In Britain crested newt habitat is most commonly located in agricultural land (50% of breeding sites in the national survey were field ponds, less than 1% in reserves). This being so, the species is commonly the subject of land use conflicts and two case studies are reviewed briefly below.

Newts on a Potential Opencast Coalmine Site

Several crested newt ponds occur at a prospective 470 Ha opencast site near to Bolton, north-west England, (Fig. 10). Surveys of the ponds in three consecutive years by several individuals indicated that none of the ponds supported a strong crested newt population (the highest count by torchlight as a single pond was 16 adult newts and the median count was one). Nevertheless, opponents of the mine argued that the site as a whole was important as a newt habitat and this was one of the issues at a public inquiry into British Coal's application to mine the site. After the planning inquiry, the Secretary of State for the Environment ruled that part of the proposed mine area should be entirely protected from opencasting. The habitat in this protected area should be improved and newts translocated to it from the area to be mined. This has resulted in British Coal's decision to create three conservation areas harbouring about 75% of the known newts; the locations of two of these areas are shown in Fig. 10. It is planned to translocate as many as possible of the remaining newts into these areas. The extra numbers will be accommodated through upgraded terrestrial habitat. The consequence of translocation is poorly understood. In particular we cannot predict the reactions of newts introduced into an existing newt population. In response to this difficulty British Coal will fund experiments into translocation.

Newts on a Proposed Building Site

The impact of conservation legislation can be seen on a smaller scale in the example of a housing development at Sutton Park in Cambridgeshire involving plans for 85 houses on an agricultural site which included a crested newt breeding pond. The probability of objections from conservation organisations lead to the modification of the plans even before they were submitted to the planning authorities. In the revised plans the pond was retained together with surrounding terrestrial habitat and the number of houses reduced from 85 to 75. In this and similar undramatic examples legislation is having a significant impact upon the ability of the species to survive in the British countryside outside special reserves.

Fig. 10: Diagram of part of the projected "Lomax" opencast coal site showing the dispersion of ponds. (● Ponds with crested newts, ○ ponds with other amphibia).

Common Species

The conservation status of the four species other than the natterjack toad and crested newt is even more difficult, since their habitats are not protected. Nevertheless, public opinion or local planning policy does increasingly lead to their conservation. An example is provided below.

Toad Site Bisected by a New Road

The toad breeding site referred to above was the subject of development plans in 1987, a new road being proposed in an east - west direction, cutting through part of the pond and separating the pond from habitat to the north (Fig. 11). A knowledge of the pattern of adult dispersion, influx routes and metamorph dispersal patterns, taken together with the assumed effects of barriers to movement enabled the construction of a predictive map of habitat quality zones around the pond (Fig. 12). This in turn was used to estimate the likely damage to the population by construction of the road and facilitated the design of tunnel and barrier systems which might be used to ameliorate the impact of the road. A knowledge of the temporal pattern of movement (Fig. 9) was significant in determining the periods when construction work at the site should be avoided. The details of the proposed tunnels and barriers are described elsewhere (OLDHAM 1989).

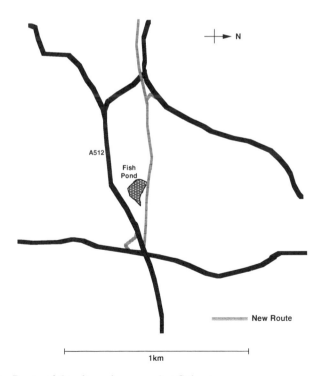

Fig. 11: Route of the planned new road at Coleorton.

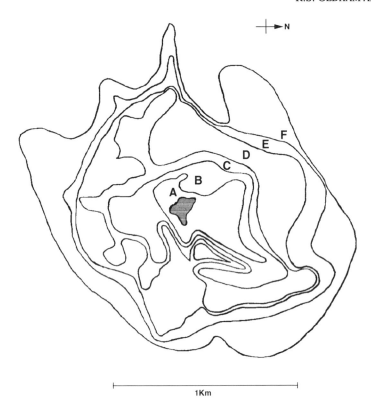

Fig. 12: Habitat zones around the Fish Pond, Coleorton. Grades A (high) to F (low) allocated on
         the basis of distance from the pond, obstacles to toad movement, known dispersion
         patterns and habitat qualtity.

Translocation as a Solution to Land Use Conflicts

Of the practical conservation measures, translocation is perhaps the most problematic
and contentious. There are practical difficulties, since only the adult component of the
population can be readily captured (by use of a perimeter fence and traps around the
breeding pond). In British populations of *Bufo bufo* we have studied, we estimate that
adults constitute only 15% of the total spring population. A perimeter fence would need to
be retained for several years to catch a significantly higher proportion of the population.
Translocation also involves serious biological problems in selection of a suitable host site.
Sites which provide ideal habitat are likely to contain strong populations already and the
effects of introducing new individuals may cause over-use of resources and, therefore,
damage the habitat, or cause increased competition and mortality in both the host and the
introduced populations. Conversely, sites which do not contain strong populations are liable
to be substandard and again unable to accommodate introduced animals, unless upgraded

well beforehand. There are also political objections to translocation which may be seen as a 'soft option', i.e. developers ridding themselves of an awkward problem rather than incorporating conservation measures into their development plans. Newly created habitats may provide a means of overcoming these problems but such habitats would need to be established several years beforehand.

The Efficacy of Amphibian Conservation

At the start of this paper we identified several essential components of successful species conservation; a knowledge of the national distribution and status, of dispersal patterns and habitat requirements, together with an appropriate public attitude and legislative framework.

Documentation of the distribution and status of amphibia at present is estimated to encompass only about 6% of existing sites in mainland Britain. Yet this is the result of intensive volunteer effort over a three year period. Clearly the effort needs to be maintained, the survey extended and existing sites periodically re-surveyed. Alternatively, a more selective monitoring scheme might be initiated (SWAN & OLDHAM 1990). Knowledge of the habitat requirements of the six British species is steadily improving and new genetic and radio tracking techniques, we are currently employing in collaboration with Leicester University promise to facilitate the determination of dispersal patterns. Such information is of little value, however, in the absence of a suitable public attitude and legislative framework. We have made reference to the benefits of involving the public in the national survey. The concomitant press coverage has ensured that knowledge of the scheme is widely disseminated and has helped to increase public awareness and sympathy for species protection. The most recent relevant legislation, the Wildlife and Countryside Act (1981), requires prospective developers to have regard to the protection of two of the British species, the natterjack toad and the crested newt. The other four species are not formally protected, but an increasing awareness of conservation is commonly manifested in the provision of practical protection measures (e.g. LANGTON 1989 for toads and OLDHAM 1989 for toads and frogs). The existing planning framework in England and Wales is illustrated in Fig. 13.

Planned developments are submitted to the relevant authorities (county or district level) (Fig. 13; (1)). If the development would affect a designated site; an N.N.R. or S.S.S.I. or a protected species (natterjack toad or crested newt), the planning authority is under obligation to consult the N.C.C. (2) who may then negotiate for habitat safeguards with the developer (3). This may lead to amended plans, withdrawal of the plans or there may be no agreement. For non-designated sites written information on the planned development is posted at the site, neighbours are notified and reports appear in the press (4). Written objections to the development can then be made and it is at this stage that the existence of significant amphibian populations may be brought to the attention of the planning authorities (5), sometimes by an authority's ecologist, or private conservation groups such as the County Conservation Trusts, by an N.C.C. officer or by individuals. The planning committee subsequently considers the application, together with any objections (6). They may approve the development, refuse it, or grant modified approval. The developer can appeal against a refusal or modified approval, in which case a public inquiry may be instituted. A decision by the Secretary of State for the Environment, who is advised by the inquiry inspector, follows. From the conservation viewpoint there are several points of weakness in this concatenation, perhaps the most significant of which are a lack of information (5), an unsympathetic planning committee (6), and insufficient funding to compensate the developer (3). These weaknesses can be addressed, respectively, by increased survey effort, increased public sympathy and increased conservation funding. Those sites where successful conservation measures have been undertaken are those in which appropriate local knowledge was already available, or in which there was time to undertake prior investigations, coupled with strong public sympathy. One of the most encouraging, if undramatic, trends in recent years is the tendency for developers to commission "land sifts" **prior** to submitting application for planning permission, the object being to identify and avoid potentially contentious sites.

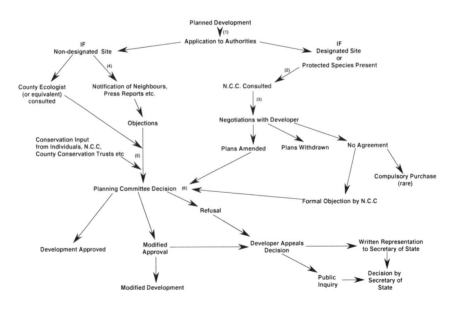

Fig. 13: Steps involved in the planning process in England and Wales as applicated to sites of conservation interest (numbers refer to comments in the text).

Zusammenfassung

Artenschutz schließt eine Einschätzung der Verteilung und des Status ein, die Kenntnis von Habitatausstattung und Ausmaß der Ausbreitung sowie die Angabe von praktischen Maßnahmen, mit denen ein Überleben gesichert werden kann. Hinzu kommen ein passender gesetzlicher Rahmen und die Einstellung der Öffentlichkeit.

Die Abschätzung der Verteilung und Häufigkeit auf der Britischen Insel schloß eine landesweite Untersuchung ein, deren Methoden, Ergebnisse und Grenzen behandelt werden. Das Wissen über die Habitatausstattung und die Verteilungsmuster - beide notwendig für Schutzmaßnahmen - können auf Populationsebene durch Fragebogenaktionen und durch Feldstudien gewonnen werden. Unsere Arbeit über den Kamm-Molch (*Triturus cristatus*) und die Erdkröte (*Bufo bufo*) dient als Beispiel. Praktische Schutzmaßnahmen können die Einrichtung von Schutzgebieten, die Anlage von Tunnels, Lenkung der Wanderung, Umsiedlung von Populationen und ein Habitatmanagement einschließen. Diese Möglichkeiten werden am Beispiel der beiden erwähnten Arten diskutiert. Die Wechselwirkung und der Einfluß von Gesetzgebung und oöffentlicher Meinung werden diskutiert. Beide sind für den Schutz ebenso wichtig wie biologische Aspekte. Der Amphiebienschutz in Großbritanien muß noch viel leisten, aber der Fortschritt ist nicht vollends entmutigend.

Acknowledgements

We are very grateful to Dr. Seitz and the sponsors for the opportunity to present this paper at the conference in Mainz; to the N.C.C. for the financial support of the National Amphibian Survey; to Leicestershire County Council and British Coal for permission to use material from studies they have funded; and to A.S.Cooke, G.F.Collings, D.Wightman and our colleagues in the Department of Applied Biology, especially D.J.Bullock and J.F.Flanagan, for help with various aspects of the work.

References

BEEBEE T.J.C. (1988): How not to save a species. New Sci. 120 (1640): 64-65.

COOKE A.S. & OLDHAM R.S. (1988): Protecting amphibians. New Sci. 120 (1640): 88.

HER MAJESTY'S STATIONERY OFFICE (1981): Wildlife and Countryside Act. H.M.S.O.

LANGTON T.E.S. (1987): Tunnels and temperature: results from a study of a drift fence and tunnel system for amphibians at Henley-on-Thames, Buckinghamshire, England. In: LANGTON T.E.S (Ed.) Amphibians and Roads. Proc. Toad Tunnel Conf; Rendsburg. Aco Polymer Products Ltd; Shefford, Beds.

NATURE CONSERVANCY COUNCIL (1989): Guidelines for selection of biological S.S.S.I.s. N.C.C. Peterborough; U.K.

OLDHAM R.S. (1985): Toad dispersal in agricultural habitats. Bull. Brit. Ecol. Soc. 16: 211-215.

OLDHAM R.S. (1989): Potential tunnel systems at road developments in England. In: LANGTON T.E.S (Ed.) Amphibians and roads. Proc. Toad Tunnel Conf.; Rendsburg. Aco Polymer Products Ltd; Shefford, Beds.

OLDHAM R.S. & NICHOLSON M. (1986): Status and ecology of the warty newt (*Triturus cristatus*). Contract report to the Nature Conservancy Council, HF3/05/123.

SWAN M.J.S & OLDHAM R.S. (1989): Amphibian communities. Ibid.; HF3/03/332.

SWAN M.J.S & OLDHAM R.S. (1990): Herptile sites. Ibid.; F72-15-04.

Species Conservation: A Population-Biological Approach
A. SEITZ & V. LOESCHCKE (eds.) © 1991 Birkhäuser Verlag, Basel

Pollination: An Integrating Factor of Biocenoses

S. Vogel [1] **and C. Westerkamp** [2]
[1] Institut für Spezielle Botanik, Universität Mainz, Saarstr. 21, D-6500 Mainz, Germany
[2] Botanisches Institut, Universität Bonn, Meckenheimer Allee 170, D-5300 Bonn, Germany

Abstract

A survey is given of the ecological constraints which affect the process of pollen transfer - and hence the gene flow - within a biocenosis. Wind pollination (anemophily) plays a dominant role in species-poor communities only. The quantity of zoophilous species increases equator-ward to up to 100 % and so does the degree of integration in animal-plant interactions. Biotic pollination is pinpointed. Manifoldness and specifity of methods reduce pollen waste and mispollinations. Saturated ecosystems dispose of a complete set of pollination syndromes and the respective pollinator guilds, narrow niche widths, a high percentage of eutropic flowers, shorter flowering times, and a temporal and spatial partitioning of floral resources. In pioneer communities generalists (allotropic flowers) or autogamy prevail. Flowering times are extended and there is less biotic integration. - Usually, floral contrivances favour allogamy. Genetic heterogeneity of potential sexual partners is effected by seed dispersal while gene flow results from pollen transfer. The mean range of the latter and thus the sizes of an individual's interaction sphere within a population (neighbourhood) are species-specific and depend on, e.g., gregariousness; efficiency of the pollination apparatus; nectar quantity; kind of pollinator, its flying range, and flight direction as well as terrain features. Usually, it amounts to a few meters only. Anemophilic and generalist species suffer from a more intense pollen waste than do specialists, but compensate for it by high pollen/ovule ratios. Habitually sparse species are pollinated by trapliners or benefit synergistically from the presence of a similar dominating species. Synergism by vicariating flowering times secures the continuity of a common pool of polylectic pollinators. An oligolectic pollinator's dependency of certain flowers is often more obligatory than vice versa. Establishment of species with specialized flowers requires additional prerequisites for the existence of respective oligolectic pollinators: nectar and pollen in due proportion, nesting facilities, foodplants or prey for the larvae, climatic conditions, etc. These dependencies and minimum population sizes are critical factors for the survival of a biocenosis.

Introduction

Plants predominate in terrestrial ecosystems, be it with a few species only or in a rich mosaic. In Costa Rica, for example, WHITMORE *et al.* (1985) counted 233 vascular plant species on just 100 m² of tropical rain forest, 365 were found by GENTRY & DODSON (1987) on a plot of 0,1 ha in Ecuador, and 243 (cit. after GENTRY 1988) in an equal area of Sarawak (Borneo). Not assorted like museum pieces but engaged in a web of interactions, they compose biocenoses. It is only now that we are beginning to understand the importance of direct and indirect interrelations of animals and plants. Phytophages, ants and soil arthropods play the main role as antagonists or symbionts in the vegetative sphere of plants. Here we will envisage animal-aided reproduction, and pollination of flowers in particular.

Plants are rooted and are hence in need of foreign vectors for their exchange of gametes or dissemination. This "borrowed mobility" (SCHREMMER 1969) is obtained from wind and animals, in propagation as in pollination. Each plant species has its own way of achieving these two purposes. There are as many strategies as there are taxa.

Diversity of Pollination Modes and their Coexistence in Ecosystems

Wind as a means of transportation plays a major role in species-poor biocenoses only. Wind dispersal loses its significance as one approaches the equator. In tropical rain forests, 70 - 98 % of all tree species are zoochorous, i.e., depend in their propagation on frugivorous animals (HOWE & SMALLWOOD 1982). The case is even more extreme for anemophilous plants: Wind-pollinated species abound in cooler zones and grass-savannas because wind pollination only works in communities with vast numbers of individuals of a few species. In the New World, 50 % of forest trees are zoophilic at 20° N, but almost 100 % are zoophilic in Amazonia (REGAL 1982). In alpine floras, too, biotic pollination is rife despite harsh conditions.

The more or less numerous zoophilic plant species forming a community are able to coexist because their pollen transfer is separate and directed, and manifold mechanisms counteract mix-up and misdirection. This is also connected with a great diversity of resources (trophic niches): Heterogeneous flower constructions, diverse baits (nectar, pollen, etc.), a different spatial arrangement and stratification of flowers within a biotope, seasonal and diurnal pattern of flowering times, etc. Accordingly, a multitude of flower visitors exists with different motifs and methods. The richer in species a community is, the more its niches are subdivided: For instance, by adaptive radiation of flowers corresponding to different proboscis lengths; or higher species number of pollinators through differentiation of host taxa by growth forms, soil preferences and other vegetative parameters.

Niches in terms of floral ecology may accordingly be wide or narrow. In species-rich climax communities niche width and abundance of individuals decrease, while the grade of integration increases. This means more specialized ("euphilic") flowers, more zygomorphy, blue and red floral colours as well as short flowering times. Poor communities or those of colonists, ruderal plants and annuals form the other extreme; fluctuation is too strong, lifetime too short for the formation of reciprocal adaptations and higher integration; autogamous plants and generalists predominate. The latter own "allophilic" (poorly specialized) blossoms (e.g., composites, umbellifers) open to a wide spectrum of opportunistic visitors, often combined with long flowering times. Such systems prevail in naturally inclement or unstable areas and in human surroundings (OSTLER & HARPER 1978; DEL MORAL & STANDLEY 1979, PROCTOR 1978, PARRISH & BAZZAZ 1979).

The diversity of flowers and pollination agents is neither chaotic nor continuous, but structured by more or less distinct syndromes ("biotypes") as they are classified by systems of floral ecology. Many tropical ecosystems are "saturated" with a complete set of euphilic associations; that means, all guilds of flower-visitors - partly rich in species - coexist within them; these associations consist of flies, butterflies, moths, bees, birds, and bats, and the corresponding flower-classes ("floral styles") specialized in them. Both partners create and fill a pattern of ecological vacancies largely independent of their taxonomic affiliation, a fact that leads to convergencies in widely separated floras. Introduced or migrating plants may be incorporated into foreign floras and find an equivalent pollinator there. On the other hand, the pollination modes are regionally coined by kinship-dependent special properties and random in their presence, because the particular biota became members of the biocenosis by historical fortuity.

Nevertheless, in a biotope, there is no clear connection between the species number of flowering plants and that of their pollinators. Usually, no species-to-species bond exists even in eutropic blossoms; instead an entire guild of visitors belongs to each floral style with species exchangeable. In a Costa Rican dry forest, e.g., 330 anthophilous bee and wasp species where found pollinating 168 species from 52 families of plants (HEITHAUS 1979). In a fallow limestone grassland in the Kaiserstuhl area (SW Germany) 128 bee and 56 butterfly species visited 71 plant species (KRATOCHWIL 1988). Sometimes, a large number of plant species shares but a few pollinators. In the neotropics, one species of flower bat serves, on average, 17 bat-flower species (compiled from DOBAT & PEIKERT-HOLLE 1985). In Madagascar, hundreds of ornithophilous species are attended by two species each of Nectariniidae and Philepittidae, and one *Zosterops*. The Galapagos Islands possess 450 species of phanerogams (GOOD 1974), but only one bee, a species of *Xylocopa* (LINSLEY 1966). That it may be the other way round is shown by the bee fauna of a barren field in an urban area of 1,5 km² in Sapporo (Hokkaido, Japan) with about 25 host species but more than 100 bee species (SAKAGAMI & FUKUDA 1973). It is obvious that the diversity of these bees - mainly polylectic Halictids - is determined by factors other than just flowers. Broad food niche overlaps are also observed in subarctic communities where 8 - 9 bumblebee species may temporally concentrate on the same set of three or less host species (BERGWALL 1979, LUNDBERG & RANTA 1980, RANTA *et al.* 1981).

Populational Aspects of Pollen Flow and Directionality

The pollination systems of a biocenosis may be studied from different perspectives, e.g.:

- their taxonomically dependent structural peculiarities
- their energetical role in the food cycle,
- their role in the population dynamics of particular kinships and partners.

Only the latter is taken into consideration here. To survive, a species must first reproduce at all and then sufficiently so; furthermore it will only persist if its populations on the long range escape from homozygosity, immobility and depression. To this end fertilization is necessary: allogamy if possible, i.e. amphimictic recombination. This principle is evident from all floral arrangements which prevent self-pollination. Depression (degeneration) as a result of inbreeding may emerge after only a few generations, as is well known from crop plants. There is hardly any data on feral populations, unfortunately. After all, some species in Central Europe reproduce almost exclusively by vegetative means without apparent draw-back, e.g., *Dentaria bulbifera*, *Ranunculus ficaria* or *Lysimachia nummularia*.

For allogamy to take place, there is a need for neighbouring conspecifics which are genetically divergent. This intermingling is brought about by seed dispersal. Geneflow and recombination, however, are achieved by pollen transfer which depends on several external and internal factors (LEVIN 1978). Each species shows its own demographic behaviour and has to be investigated individually to answer questions such as:

(1) Which proportion of pollen from an individual reaches the stigmata of conspecifics? Respectively: Which proportion of its ovules is fertilized (fecundity rate)?

(2) What is the range of pollen transport, and how many individuals are actually reached?

The efficiency of pollen dispersal (1) depends on the mechanics of pollen presentation, visitor behaviour and variables such as density of populations, competition with other plants, intensity of flower visitation or wind, respectively. Generalist visitors spoil more pollen and hence are less effective than specialists, to which a flower is adapted. Constant heavy losses are compensated by high pollen/ovule ratios. In andromonoecious taxa this depends on the proportion of staminate flowers, in dioecious taxa on that of staminate or hermaphrodite individuals. Their populations, by undergoing selection according to local circumstances, may genotypically regulate the amount of pollen available (ASSOUAD *et al.*, 1978; DOMMEE *et al.*, 1978), or adapt their sex rate by pollen tube competition in the style (MULCAHY, 1967). As a fail-safe mechanism to secure reproduction, phenotypically conditioned autogamous or kleistogamous morphs (e.g., *Impatiens* spp.; SCHEMSKE 1978, WALLER 1980) are developed by many taxa.

Most flowers are prepared for repeated visits and a particular vector may simultaneously deposit pollen of more than one conspecific donor. Multiple paternity is typical for plants, i.e., the seeds of one fruit usually have several fathers, thus additional heterogeneity is produced in the progeny. In radish (*Raphanus sativus*), ELLSTRAND (1984), by way of gene-markers, was able to demonstrate 4 parents in 8 % and at least 3 parents in 20 % of the pods. On the average, multipaternal fruits (in *Costus allenii*) resulted in a higher seed set and in stronger progeny than those of a sole father (SCHEMSKE & PAUTLER 1984).

The reach of a particular pollen donor (2) in meters is known only in a few cases. People suffering from hayfever know that wind pollen is distributed farther than pollen adapted to animal vectors. Whether this pollen ever reaches its destination is another problem. The first question to be posed in animal-pollinated flowers is how far their visitors fly at all. Certainly, vertebrates and some Lepidoptera are able to mediate genetic drift between different populations; in bees (Euglossini, *Xylocopa*), too, flight distances have been demonstrated to exceed 20 km (JANZEN 1971). The ecological range of foraging honeybees includes over 300 km^2 for any given colony, tapering gradually from the nest (ROUBIK 1989). In solitary bees (*Osmia, Megachile, Andrena*) ranges of 130 - 800 m were measured (WESTRICH 1989). These data reflect reality only if flights are not erratic and vectors do not return to flowers just visited. Indeed, bees fly more or less rectilinear, bumblebees fly mostly into the wind. Linear ground structures like crests or rivers canalize, too. Biotic constraints also help to lengthen foraging flights, e.g., pressure exerted by enemies and competitors or relatively small nectar portions, which are worth only a short stop at each flower and necessitate visits to a larger number of flowers (FRANKIE 1976).

It does not follow necessarily from these observations that pollen really reaches thus far. The fate of single pollen grains was traced by radioactive or fluorescent marking. After visiting a *Phlox* flower, a *Colias* butterfly carried about 10% of the flower's pollen on its proboscis, 10 - 17 % of which remained at the second flower, i.e., about 1 % per flower (LEVIN & BERUBE 1972, cit. after RICHARDS 1986). In the case of an *Erythronium,* 50 % of the initial pollen mass was deposited at the next flower, and less than 1 % arrived at the eighth following flower. Bumblebees removed a mean of 62.1 % of the pollen presented by the donors and delivered a mean of 0.52 % of the removed pollen to the stigmas of other individuals during a bout (THOMSON & PLOWRIGHT 1980, THOMSON & THOMSON 1989). In multi-flowered inflorescences, most often geitonogamy results which is equivalent to autogamy. In *Primula* species with dimorphic pollen, the fate of grains on neighbouring plants' stigmas is directly computable. Most receivers were at a distance of less than 2 m, the remotest at 16 m (RICHARDS 1986). In a *Delphinium*, the mean distance was 1 m (PRICE & WASER 1979), in certain violets (*Viola*) 50 cm, on average (BEATTIE 1978).

The effective radius of a pollen donor is thus unexpectedly small. The pollen flow curve (measured as quantity per distance) is sloping steeply, i.e., leptocurtic. As a rule, an individual does not reach the boundaries of its population. It was attempted to determine the limited actual area of panmixis or breeding unit, termed neighbourhood. The more crowded (patchy) the individuals, the smaller the neighbourhood and the slower the gene flow. Neighbourhood sizes (data compiled from RICHARDS 1986) were estimated at 29 500 m² with 1 - 3200 trees in anemophilous pines, at 1600 m² with 230 individuals in poplars. In entomophilous herbs like violets or primroses, on the other hand, it amounted to 19 - 57 m² with 167 - 574 plants or 20 - 30 m² with 5 - 200 individuals, respectively. The "paternity pool" sizes, i.e., the mean number of potential fathers a given individual's progeny has, is accordingly small in herbs and large in wind-pollinated trees (LEVIN 1988). With several groups of animals of different ranges participating in the pollination of a plant species, it has neighbourhoods of various dimensions; in certain alpine *Senecio* species of Colorado, growing in patches, e.g., they were estimated to comprise 23 individuals and an area of 0.75 m² on average as far as bumblebees were concerned, but 6000 plants on 200 m², taking vagrant butterflies into account (SCHMITT 1980).

An exception from leptocurtic pollen dispersal is shown in species using trap-lining pollinators (see, e.g., LINHART 1973). This strategy is acted upon by several nectar-rich plants in tropical forests. Typically, their individuals grow solitarily at large distances and are visited in turn by far-flying bees or birds daily. Some trees, too, belong here, e.g., *Tabebuia* (Bignoniaceae) with strongly synchronous and mass-flowering crowns widely visible above the canopy, originally perhaps including temperate rosaceous fruit-trees. In these cases, neighbourhood and population sizes usually escape evaluation.

Other plants are scattered by nature without using trap-liners. As singles, they are neither attractive nor sufficiently rewarding to tie pollinators. Their way of survival is synergism with a co-occurring "magnet-species", i.e., they share their pollinator with another similar but more frequent species though at the expense of mixed pollination. In a test case, the dominating favorite *Hieracium aurantiacum* significantly increased visitation rates in the rare *H. florentinum* (THOMSON 1978). Many terrestrial orchids, e.g., *Orchis* spp., parasitically take advantage of this kind of participation, being mimics without any reward.

Certain long-flowering species or those just blooming at a critical time ("keystone species") play a role in overcoming temporal bottlenecks. The members of a biocenosis collectively profit from the persistance especially of polylectic pollinators in the system, i.e., if starvation or emigration for a want of food are prevented. Forming seasonal niches, floral hosts with vicarious flowering times alternate during the course of the year, e.g., bird flowers of a Costa Rican forest (STILES 1979). Mixed pollinations are reduced to a minimum by this time-partitioning - another way of synergism. Consequently, the subsistence of pollinator populations more or less represents a matter of risk for communities.

Degrees of Interdependency among Flowers and Pollen Vectors

While plants pollinated by guilds or generalists tolerate the exchangeability of their visitor species, specialists are obligately adjusted to a particular vector. Members of the genus *Ficus* carried their pollinators along with them everywhere while speciating and propagating, as fig wasps complete their total life cycle within their host and thus lend them independency. But usually, a number of diverse ambient factors is needed to establish the particular pollinators of a colonizing host. An example for climatic factors is provided by *Wisteria sinensis*; introduced into Central Europe as a garden plant, it (partially) sets fruit merely in the Rhine valley, as carpenter bees (*Xylocopa violacea*), their sole pollinators, thrive here only.

Often, the occurrence of the host is of more vital importance for their pollinators than vice versa. This applies to, e.g., oligolectic solitary bees; their bonds are generally based on specific kinds of pollen, without which their larvae cannot succeed (BOHART & YOUSSEF 1976). Their host flowers, however, can manage with substitute pollinators or are even generalists. Commonly, polylectic bees fill the gap which are able to cope with very different kinds of flowers. They, e.g., honeybees, temporally "major" on a rewarding source, i.e., they become flower constant spontaneously from experience instead of instinctively. There are no flower species indeed, which are especially adapted to the honeybee, *Apis mellifera* (WESTERKAMP 1987), with the exception perhaps of some mass-flowering species of early spring, e.g., *Prunus*, *Galanthus* or *Leucojum* (WESTRICH 1989). The actual anthropogenous excess of honeybees is unnatural and noxious to communities, because they extrude solitary bees from their hosts but do not even fairly take over their pollinating service.

About 400 species of 3200 Central European phanerogams nourish bees, and 21 genera are sole hosts of a bee species. Of 380 non-parasitic bee species in Germany, 116 (ca. 30 %) are known to be oligolectic, i.e., dependent on at least one particular plant family, 48 of them on a single genus (WESTRICH 1989). Even in the apifauna of anthropogenous biotopes like ruderal sites in Freiburg (SW Germany), up to 50 % of specialists (especially on Asteraceae, Fabaceae, and Apiaceae) were recorded (KRATOCHWIL & KLATT 1989).

Ambiental Prerequisites for the Maintenance of Pollination Systems

Bees are in need of pollen and nectar, and both must be available in due proportion in a place. Most nectar flowers are inadequate pollen sources. Mere pollen flowers then gain a key role as they compensate for it. If bees are specialized on the pollen of pollen-poor nectar flowers, e.g., visitors of *Campanula* (WESTRICH 1989), then the host population must be accordingly large. Many anthophorid bees, especially *Centris*, which belong to the most important pollinators of tropical America, need fatty oils as larval provision besides pollen and nectar (VOGEL 1989). This oil is provided by oil-flowers only, in particular those of the

Malpighiaceae. They must be available wherever *Centris* lives, and so are an indirect prerequisite to pollination of numerous nectar and pollen flowers of the same biotope. Oil flowers, e.g., *Calceolaria*, for their part can only maintain their pollinators, if these find pollen and nectar flowers in the vicinity as well. In the Sonoran desert, e.g., the *Centris*-pollinated oil-yielding *Krameria* synchronizes its short flowering time fairly well with that of *Prosopis*, *Larrea* and *Cercidium*, the pollen and nectar hosts of its visitors (SIMPSON & NEFF 1987).

The nesting habits of solitary bees call for definite materials and soils - an indispensable prerequisite, too, for reproduction of their host plants. Many of our ground-nesting species make their burrows only in fine-grained soils sloping southward; others like *Anthidium* build their cells from plant wool of certain *Cirsium*, *Verbascum* or *Stachys* species. Others again nest in hollow stems provided by *Verbascum* or umbellifers, or (e.g., *Osmia* spp.) in snail-shells (*Cepaea, Helicella*), which must be available in sufficient numbers. To offer a tropical example, the flower bat *Eonycteris spelaea* may be cited (START & MARSHALL 1976), the most important pollinator of Malayan bat flowers. In the daytime they gather in a few caves of Central Malacca, from where they fly long distances every night. A destruction of their sleeping quarters would cut off reproduction of several host plant species in the whole country, including fruit trees and mangroves.

The visitors of psychophilous and sphingophilous flowers, i.e., butterflies and hawkmoths, are dependent as immatures on further foodplants which are mostly different from their floral hosts. In Europe, they are pollinators of, e.g., honeysuckle (*Lonicera* spp.), catchfly (*Silene* spp.) or soap-wort (*Saponaria*); their caterpillars, on the other hand, feed on bedstraw (*Galium*), lesser bindweed (*Convolvulus*) or privet (*Ligustrum*), respectively. The caterpillars of *Maculinea nausithoos* and *M. teleius* (Lepidoptera) in their first stages feed on burnet (*Sanguisorba officinalis*) while mining its flower heads, but later on they depend on the presence of an animal host, the ant *Myrmica rubra*, upon which they parasitize as symphiles. The emerging adults then return to *Sanguisorba* to suck the floral nectar and to deposit eggs (ELMES & THOMAS 1987; K. FIEDLER, pers. comm.).

Flower visiting hoverflies feed on aphids in their larval stage; the bee flies (Bombyliidae), also important flower visitors, are brood parasites of true bees. Pollinators of deceptive flowers are in need of their proper hosts or substratum, e.g., warm-blooded animals or birds' nests for some small flies, or genuine mushrooms for pollen-transmitting fungus gnats, etc.

How subtle dependencies can be is illustrated by the hummingbird, *Archilochus colubris*, which on its return from southern winter quarters to its Canadian breeding grounds initially feeds on tree sap as a substitute for its nectar sources, ornithophilous flowers which are not yet in anthesis. The birds are thus dependent, at that time, on the sapsucker *Sphyrapicus varius*, a specialized woodpecker which drills the sapholes (MILLER & NERO 1983).

Some Consequences for Nature Conservation

In conclusion, there are inferences of importance for nature conservation, some of which, of course, have long been recognized. Minimum population sizes must be warranted for either partner of a pollination system. This is essential particularly with specialists while generalists are better buffered because of the wider potential resource base (TEPEDINO 1979). Besides their bilateral floral relations, all other preconditions of existence must not be neglected including further organisms. We must admit that hardly anything is known on critical population sizes. For many years, in the Mainz Botanical Garden, about 20 specimens of *Lysimachia vulgaris* sufficed to maintain a population of its oligolectic pollinator, *Macropis labiata*. Artificially isolated individuals are prone to disappearance without progeny. When scattered in an altered environment they are no longer approached or accepted as a rewarding source by their potential visitors. No search image nor flower constancy will form. The rarer a plant species is, the worse this danger becomes. JANZEN (1986) hit the nail on the head when speaking of "living deads", having in mind single lone trees which are left over when clear felling a virgin forest.

In a landscape with ephemeral mass-flowering mono-cultures, feral flower-insects cannot survive; after cessation of bloom they will starve to death or abscond. Therefore, it is essential to conserve field edges, fallows and limestone grasslands as "make-shift biotopes".

Apiculture has to be kept out of nature reserves; possibilities for nesting of wild bees have to be improved on purpose, if indicated.

Survival of pollinators is clearly an essential factor for the survival of an entire biocenosis.

Zusammenfassung

Es wird ein Überblick über die ökologischen Bedingungen gegeben, denen die Prozesse der Pollenübertragung - und damit der Genfluß - in einer Biozönose unterworfen sind. Windbestäubung (Anemophilie) spielt nur in artenarmen Pflanzengesellschaften eine Rolle. Die Zahl der zoophilen Arten nimmt äquatorwärts bis gegen 100% zu und damit der Integrationsgrad der tier-pflanzlichen Beziehungen. Biotischer Pollentransport ist zielgerichtet; Pluralismus und Spezifität der Methoden vermindern Pollenverschleiß und Fehlbestäubungen. Saturierte Lebensgemeinschaften verfügen über ein vollständiges Spektrum von Blütenbiotypen und entsprechenden Bestäuber-Gilden, geringe Nischenbreiten, einen hohen Anteil eutroper Blüten, kürzere Blühzeiten, temporale und räumliche Partitionierung des Blütenangebots. In Pioniergesellschaften dominieren Generalisten (allotrope Blüten) oder Autogamie. Die Blühzeiten sind länger; die biotische Integration ist geringer.- Im allgemeinen folgen die Blüteneinrichtungen dem Imperativ der Fremdbefruchtung (Allogamie). Die genetische Heterogenität potentieller Sexualpartner wird durch die Samenverbreitung, der Genfluß selbst durch Pollentransfer bewirkt. Dessen mittlere Reichweite und damit die Größe der Kommunikationssphären innerhalb einer Population (neighbourhood) ist artspezifisch und hängt u.a. von der Bestandsdichte, Effektivität des Pollinationsapparates, Nektarmenge, Art und Flugweiten sowie Flugrichtung des Überträgers und von der Geländestruktur ab. Meist beträgt sie nur wenige Meter. Anemophile und Generalisten erleiden stärkeren Pollenverschleiß als Spezialisten, was durch hohe Pollen/Ovula-Relationen kompensiert wird. Natürlicherweise vereinzelt wachsende Arten befolgen die Trapliner-Strategie oder profitieren synergistisch von einer ähnlichen dominierenden Art. Synergismus in Form vikariierender Blütezeiten sichert Kontinuität eines gemeinsamen polylektischen Bestäuberkreises. Die Abhängigkeit oligolektischer Bestäuber von ihren Spezialwirten ist oft strenger als umgekehrt. Bei der Ansiedlung solcher Pflanzenarten müssen jedoch zusätzliche Existenzbedingungen für deren Bestäuber erfüllt sein: Nektar- und Pollenangebot in bestimmtem Proporz, Nistgelegenheit, Nährpflanzen und Beute für die Larvenstadien, klimatische Ansprüche u.a.. Diese Abhängigkeiten und Mindestpopulationsgrößen sind kritische Faktoren für die Überlebensfähigkeit einer Biozönose.

References

ASSOUAD, M.W., DOMMEE, B., LUMARET, R., VALDEYRON, G. (1978): Reproductive capacities in the sexual forms of the gynodioecious species *Thymus vulgaris* L..- Bot. J. Linn. Soc. 77: 29-39

BEATTIE, A.J. (1978): Plant-animal interactions affecting gene flow in *Viola*.- pp. 151 - 164 in: RICHARDS, AJ (ed.): The pollination of flowers by insects.- Linn. Soc. Symp. Ser. 6. London: Academic Press

BERGWALL, H.E. (1970): Ekologiska iakttagelser över några humlearter (*Bombus* Latr.) vid Staloluokta inom Padjelanta nationalpark, Lule, Lappmark.- Entomol. Tidskr. (Stockholm) 91: 3-23

BOHART, G.E., YOUSSEF, N.N. (1976): The biology and behavior of *Evylaeus galpinsiae* Cockerell (Hymenoptera: Halictidae).- Wasmann J. Biol. 34: 185 - 234

DEL MORAL, R., STANDLEY, L.A. (1979): Pollination of angiosperms in contrasting coniferous forests.- Am. J. Bot. 66: 26 - 35

DOBAT, K., PEIKERT-HOLLE, T. (1985): Blüten und Fledermäuse. Bestäubung durch Fledermäuse und Flughunde (Chiropterophilie).- Senckenberg-Buch 60. Frankfurt: Kramer.

DOMMEE, B., ASSOUAD, R. & VALDEYRON, G. (1978): Natural selection and gynodioecy in *Thymus vulgaris* L.- Bot. J. Linn. Soc. 77: 17-28

ELLSTRAND, N.C. (1984): Multiple paternity within the fruits of the wild radish, *Raphanus sativus*.- Am. Nat. 123: 819 - 828

ELMES, G., THOMAS, J.A. (1987): (on *Maculinea*) in GEIGER, W. (ed.): Tagfalter und ihre Lebensräume.- Basel: Schweizer Bund f. Naturschutz

FRANKIE, G.W. (1976): Pollination of widely dispersed trees by animals in Central America, with special emphasis on bee pollination systems.- pp. 151 - 159 in: BURLEY, J., STILES, B.T. (eds.): Tropical trees - variation, breeding and conservation.- Linn. Soc. Symp. Ser. 2. London: Academic Press

GENTRY, A.H. (1988): Changes in plant community diversity and floristic composition on environmental and geographical gradients.- Ann. Missouri Bot. Gard. 75: 1 - 34

GENTRY, A.H., DODSON, C. (1987): Contribution of non-trees to species richness of a tropical rain forest.- Biotropica 19: 149 - 156

GOOD, R. (1974): The geography of flowering plants.- London

HEITHAUS, E.R. (1979): Flower visitation records and resource overlap of bees and wasps in northwest Costa Rica.- Brenesia 16: 9 - 52

HOWE, H.F., SMALLWOOD, J. (1982): Ecology of seed dispersal.- Ann. Rev. Ecol. Syst. 13: 201-228

JANZEN, D.H. (1971): Euglossine bees as long-distance pollinators of tropical plants.- Science 171: 203 - 205

JANZEN, D.H. (1986): The future of tropical ecology.- Ann. Rev. Ecol. Syst. 17: 305 - 324

KRATOCHWIL, A. (1988): Co-phenology of plants and anthophilous insects - a historical area-geographical interpretation.- Entomol. Gen. 13: 67 - 80

KRATOCHWIL, A., KLATT, M. (1989): Apoide Hymenopteren der Stadt Freiburg i. Br. - Submediterrane Faunenelemente an Standorten von kleinräumig hoher Persistenz.- Zool. Jb. Syst. 116: 379 - 389

LEVIN, D.A. (1978): Pollinator behaviour and the breeding structure of plant populations.- pp. 133 - 150 in: RICHARDS, A.J. (ed.): The pollination of flowers by insects.- Linn. Soc. Symp. Ser. 6. London: Academic Press

LEVIN, D.A. (1988): The paternity pools of plants.- Am. Naturalist 132: 309-317

LEVIN, D.A., BERUBE, D.E. (1972): *Phlox* and *Colias*: the efficiency of a pollination system.- Evolution 26: 242 - 250

LINHART, Y.B. (1973): Ecological and behavioral determinants of pollen dispersal in hummingbird-pollinated *Heliconia*.- Am. Naturalist 107: 511-523

LINSLEY, E.G. (1966): Pollinating insects of the Galápagos Islands.- pp. 225 - 232 in: BOWMAN, R.I. (ed.): The Galápagos. Proceedings of the symposia of the Galapagos International Scientific Project. Los Angeles: Univ. Calif. Press

LUNDBERG, H., RANTA, E. (1980): Habitat and food utilization in a subarctic bumblebee community. - Oikos 35: 303-310

MILLER, R.S., NERO, R.W. (1983): Hummingbird-sapsucker associations in northern climates.- Can. J. Zool. 61: 1540 - 1546

MULCAHY, D.L. (1967): Optimal sex ratio in *Silene alba*.- Taxon 16: 280-283

OSTLER, W.K., HARPER, K.T. (1978): Floral ecology in relation to plant species diversity in the Wasatch Mts. of Utah and Idaho.- Ecology 59: 848 - 861

PARRISH, J.A., BAZZAZ, F.A. (1979): Difference in pollination niche relationships in early and successional plant communities.- Ecology 60: 597 - 610

PRICE, M.V., WASER, N.M. (1979): Pollen dispersal and optimal outcrossing in *Delphinium nelsonii*.- Nature 277: 294 - 296

PROCTOR, M.C. (1978): Insect pollination syndromes in an evolutionary and ecosystematic context.- pp. 105 - 116 in: RICHARDS, A.J. (ed.): The pollination of flowers by insects.- Linn. Soc. Symp. Ser. 6. London: Academic Press

RANTA, E., LUNDBERG, H., TERÄS, I. (1981): Patterns of resource utilization in two Fennoscandian bumblebee communities.- Oikos 36: 1 - 11

REGAL, P.J. (1982): Pollination by wind and animals: ecology of geographic patterns.- Ann. Rev. Ecol. Syst. 13: 497 - 524

RICHARDS, A.J. (1986): Plant breeding systems.- London: Unwin & Allen

ROUBIK, D.W. (1989): Ecology and natural history of tropical bees.- Cambridge: Cambridge Univ. Press

SAKAGAMI, S., FUKUDA, H. (1973): Wild bee survey at the campus of Hokkaido university.- Journ. Fac. Sci. Hokkaido Univ. VI, Zool. 19: 190 - 250

SCHEMSKE, D.W. (1978): Evolution and reproductive characteristics in *Impatiens* (Balsaminac.): The significance of cleistogamy and chasmogamy.- Ecology 59: 596 - 613

SCHEMSKE, D.W., PAUTLER, L.P. (1984): The effect of pollen composition on fitness components in a neotropical herb.- Oecologia 62: 31 - 36

SCHMITT, J. (1980): Pollinator foraging behavior and gene dispersal in *Senecio* (Compositae).- Evolution 34: 934 - 943

SCHREMMER, F. (1969): "Geborgte Beweglichkeit" bei der Bestäubung von Blütenpflanzen.- Umschau 1969 (8): 228 - 234

SIMPSON, B.B., NEFF, J.L. (1987): Pollination in the arid southwest.- Aliso 11: 417 - 440

START, A.N., MARSHALL, A.G. (1976): Nectarivorous bats as pollinators of trees in West Malaysia.- pp. 141 - 150 in: BURLEY, J., STILES, B.T. (eds.): Tropical trees - variation, breeding and conservation.- Linn. Soc. Symp. Ser. 2. London: Academic Press

STILES, F.G. (1979): El ciclo anual en una comunidad coadapta de colibríes y flores en el bosque muy húmedo de Costa Rica.- Rev. Biol. Trop. 27: 75 - 101

TEPEDINO, V.J. (1979): The importance of bees and other insect pollinators in maintaining floral species composition.- Great Basin Naturalist Memoirs 3: 139 - 150

THOMSON, J.D. (1978): Effects of stand competition on insect visitation in two species of *Hieracium.*- Am. Midl. Nat. 100: 431 - 440

THOMSON, J.D., PLOWRIGHT, R.C. (1980): Pollen carryover, nectar rewards, and pollinator behavior with special reference to *Diervillea lonicera.*- Oecologia 46: 68 - 74

THOMSON, J.D., THOMSON, B.A. (1989): Dispersal of *Erythronium grandiflorum* pollen by bumblebees: Implications for gene flow and reproductive success.- Evolution 43: 657 - 661

VOGEL, S. (1989): Fettes Öl als Lockmittel. Erforschung der ölbietenden Blumen und ihrer Bestäuber.- pp. 113 - 130 in: Akademie der Wissenschaften und der Literatur Mainz 1949-1989.- Stuttgart: Steiner Wiesbaden

WALLER, D.M. (1980): Environmental determinants of outcrossing in *Impatiens capensis* (Balsaminac.).- Evolution 34: 747 - 761

WESTERKAMP, C. (1987): Das Pollensammelverhalten der sozialen Bienen in Bezug auf die Anpassungen der Blüten.- Dissertation Mainz

WESTRICH, P. (1989): Die Wildbienen Baden-Württembergs.- Stuttgart: Ulmer

WHITMORE, T.C., PERALTA, R., BROWN, K. (1985): Total species count in a Costa Rican tropical rain forest.- J. Trop. Ecol. 1: 375 - 378

Population Biology of an Invading Tree Species - *Prunus serotina*

U. Starfinger,   Institut für Ökologie - Ökosystemforschung und Vegetationskunde -,
Technische Universität Berlin, Schmidt-Ott-Str. 1, 1000 Berlin 41, Germany

Abstract

The Black Cherry (*Prunus serotina*), a North American forest tree, has successfully spread in Central Europe. By establishing dense shrub layers in previously sparse forests, it may outcompete native plant species and become a problem for nature conservation.

Demographic studies were performed in its native range NW-Pennsylvania, and in Berlin, and the data were analyzed by means of multivariate methods. In both areas, the distribution in space and time of size classes suggests a "Oskar-behaviour" of *Prunus serotina*: the smaller trees can survive with limited height growth in the shade of the conspecific mother trees. This property is, together with other population biological characteristics, an important prerequisite for a successful colonization of new biotopes and the establishment of the species in Central Europe.

The development of a Black Cherry population in both study areas follows various phases. Without disturbance, few large Black Cherry trees prevail and little regeneration. This trend can be seen from demographic analyses both in Berlin and Pennsylvania despite their very different vegetation types. From this it is concluded that *Prunus serotina* will continue to grow in Berlin. The density, however, will be lower than now.

Introduction

Among the man-induced changes in flora and vegetation are not only extinction of species by means of destruction of sites but also the addition of new species through importation and planting. The increase in species numbers of a given region has for a long time exceeded the loss through extinction (JÄGER 1988). Whereas most of the new species are unable to naturalize and - especially - to penetrate near-natural vegetation, some can naturalize and spread and can thus become a threat for native vegetation. So nature conservation has to focus its interest not only on endangered species but also on these "dangerous species" (FUKAREK 1987).

Invasions of organisms - plant and animal species - have repeatedly been the topic of extensive studies, such as ELTON's (1958). More recently, a SCOPE (Scientific Committee on Problems of the Environment) program on the ecology of biological invasions has led to the publication of a number of books on the subject (KORNBERG & WILLIAMSON 1986, GROVES & BURDON 1986, MOONEY & DRAKE 1986, JOENJE et al. 1987, DRAKE et al. 1989, for more see MOONEY & DRAKE 1989).

In order to understand the success of an invader and to plan possible management strategies against it, it is necessary to know not only the ecosystem concerned but also the ecology of the invading species, especially its population biology.

The North American Black Cherry (*Prunus serotina*) is an example of introduced plant species that cause environmental problems. It was first introduced to Europe in 1623 (WEIN 1930). At first it was planted as an ornamental tree in parks and gardens only. From the second half of the 19th century on it was cultivated widely in the forests as an auxiliary species and for fire prevention around coniferous plantations. On and near the sites where it was planted it propagated and built up dense layers of shrubs and small trees. In the shade of these shrub layers few native species of the ground vegetation can survive and native tree species can hardly regenerate. Because of *Prunus serotina*'s adverse effects on the vegetation and of its negative influence on forestry, nature conservation and forest authorities in Berlin are trying to fight back (cf. KRAUSS et al. 1990).

Research has been done on the spread of *Prunus serotina* and its ecological consequences in the Netherlands (e.g. EIJSACKERS & OLDENKAMP 1976, TWEEL & EIJSACKERS 1987) and in Berlin (ERNST 1965, STARFINGER 1987, 1990). The objective of this study is to show how population biological traits of the species determine its role in the vegetation and its naturalization success in Europe. To do this a demographic analysis of populations in its native range in North America and in Berlin was carried out.

Methods

For the demographic analysis 75 records were taken in Berlin in the summer of 1985 and 1986, and 73 records in Pennsylvania were taken in August and September 1987. The demography of *Prunus serotina* was recorded on sample plots of 15 m x 15 m. Plots were chosen where the vegetation was homogeneous and representative for the surroundings. Together with each demographical record a phytosociological relevé was taken using the method of BRAUN-BLANQUET (1964). The large individuals of *Prunus serotina* (saplings and trees) were counted in two classes (living, T1 and dead, T2) and the circumference at breast height (CI, in cm) of each individual over 10 cm CI was measured with a measuring tape. Smaller individuals were counted in 15 subplots of 1 m^2 arranged along the diagonal of the plot. They were recorded in four stages:

S1 - seedlings germinated in the current year

S2 - with woody stem base, but not larger than S1 (mostly seedlings of the previous year)

S3 - taller than S2, but less than 1 m

S4 - taller than 1 m

The distinction between S2 and S3 is somewhat arbitrary and not in all cases exact. Under comparable conditions, however, as in the shade of the canopy, the error can be estimated to be less than 10 %.

Data Processing

The demographic data were processed on a personal computer using a package of computer programs by WILDI & ORLOCI (1983, 1988). As these programs were created for the analysis of vegetation relevés (i.e. species lists with scores), species were substituted for demographic parameters. Resemblance matrices of the normalized data were computed with the correlation coefficient. The data were then classified in a minimum variance cluster analysis. A principal component analysis was computed and its result presented in a scatter diagram.

The Study Areas

a) Pennsylvania

The Allegheny National Forest in NW Pennsylvania is the center of the most productive range of Black Cherry. Here it reaches high frequency and often dominance in mixed stands. The trees have mostly straight trunks and reach heights of up to 40 m.

The landscape is a typical plateau-landscape with rounded mountain tops and deep-cut valleys. Altitudes range from 500 to 700 m above sea level. The climate is cool and wet with a mean annual precipitation around 1200 mm and a mean annual temperature of 8 °C (COLLINS & PICKETT 1982). The main soil types are sandy loam to loam with some clay loam areas (LUTZ 1930).

The natural vegetation is transitional between temperate deciduous forest and boreal coniferous forest. Most of the sites can be classified as beech-birch-maple forest according to TAYLOR (1928). The main canopy species are American Beech (*Fagus grandifolia*), Sugar Maple (*Acer saccharum*), Black Cherry (*Prunus serotina*), Eastern Hemlock (*Tsuga canadensis*) and Yellow Birch (*Betula alleghaniensis*). The most frequent species in the ground layer are seedlings of cherry, beech and the two maple species, and of the herbaceous plants *Mitchella repens*, *Maianthemum canadense* and the ferns *Dennstaedtia punctilobula*, *Thelypteris noveboracensis* and *Dryopteris intermedia*. Light measurements showed that the relative light intensity in the shade of the beech-birch-maple forest was between 1 and 2 % of the light outside.

Most of the area was cut commercially in two phases: In the second half of the 19th century only the most valuable stems, e. g. White Pine (*Pinus strobus*) or Hemlock were removed, and between 1890 and 1930 most stands were clear-cut. Thus the vegetation now represents a successional stage that is 60 to 100 years old. Only a few hundred hectares in two nature reserves remain that have never been cut.

Human influences since the cuttings have been relatively small; in particular, no trees were planted, leaving the area to regrow naturally.

b) Berlin

Berlin is situated at 52°31' N 13°24' E. The climate is of a subcontinental type with a mean annual temperature of 9.2 °C and annual precipitation averaging around 600 mm, most of which falls during the summer months. Geomorphologically the area can be divided into two major types of landscape: the ground moranes with mostly loamy soils and the fluvioglacial valleys with poorer sandy soils. Originally most of the area - excluding open waters and peat bogs - was forested, with oak-pine communities of the phytosociological unit Quercion robori-petraeae on sand and oak-hornbeam communities of the Carpinion on loamy soils (SUKOPP *et al.* 1980).

Today, most of the 480 km^2 are densely built up, but a total of 7890 ha remains forested. Whereas flora and vegetation of the central parts of the city are considerably altered by human activities, this is less true for the forests. Most trees in the forests were planted. Among the man-induced changes to the ecosystem are a shift towards a higher percentage of coniferous trees (now 62 % of the total area) and heavy cutting during and after World War II. 42 % of the area was completely or almost completely clearcut in 1948/49 (ANONYMUS 1982).

The most frequent forest community is an oak-pine forest (*Pino-Quercetum petraeae*). The main tree species are Scots Pine (*Pinus sylvestris*) and oaks (*Quercus robur, Q. petraea*). Shrub and ground layer of the vegetation are often rather sparse and consist of canopy tree regeneration and Sorbus aucuparia, *Avenella flexuosa, Agrostis tenuis, Impatiens parviflora, Pteridium aquilinum* etc. The relative light intensity under the canopy is mostly between 5 and 15 % of the light outside.

Results

a) Pennsylvania

Fig. 1 shows the mean numbers of demographic units per record. There is a steady decline from one size class to the next one, indicating the mortality that is present in each

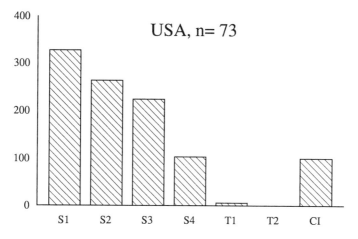

Fig. 1:   Mean numbers of demographic units per sample plot (225 m²). Records from NW-Pennsylvania. S1 - S4: Seedling classes (numbers per subplot of 15 m²), T1 - T2: Trees and saplings, living resp. dead (numbers per plot of 225m²), CI - mean circumference at breast height of T1 and T2 (cm)

size class. A mortality rate, however, can not be calculated from these data, because size is not necessarily correlated with age (HARPER 1977). A single record may look very different from the average. Some records contain few trees with large diameters and large numbers of small seedlings (S1 - S3), others contain only large seedlings (S4) or large numbers of smaller trees.

The cluster analysis of the demographic units (fig. 2) shows that the small seedlings (S1 - S3) together with the circumference of the trees (CI) form one branch of the dendrogram whereas the larger seedlings (S4) are separate from all the other demographic units. Only on a high level of dissimilarity are they linked with the number of trees (T1, T2).

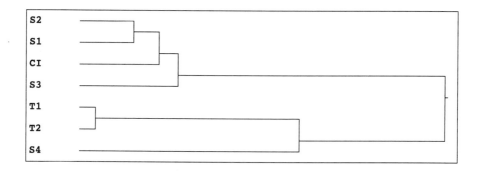

Fig. 2:   Cluster Analysis of demographic units from Pennsylvania. Resemblance measure: correlation coefficient, min. var. clustering.

This can be interpreted as a tendency of *Prunus serotina* to germinate abundantly in the shade of conspecific mother trees, the larger the tree the higher is the number of seedlings under it. Survival in the shade of the mother trees is high only up to the size class S3, the size class S4 is present where large trees are not.

With the cluster analysis of the records, similar records can be arranged in groups. 5 groups were formed with different combinations of demographic units. The groups differ mainly in the size of the trees, the number of the small seedlings and especially in the presence of larger seedlings (S4).

The spatial relation of the dots in the scatter diagram of the PCA (fig. 3) represents the similarity relation between the demographic records. The dots are arranged in such a way that they show the relationship between the groups as a trend from group 2 via groups 1 and 5 to group 3, with group 4 in the center. This trend can be interpreted as a temporal relation between the groups which then represent stages of a succession. Thus a model of the population dynamics can be deducted from the PCA (fig. 4). The population develops from a stage with a high number of large seedlings and small trees to a stage with no large seedlings and a still large number of small trees; in this stage the production of fruits begins and consequently small seedlings appear. The following are late developmental and mature stages, few trees with large diameters stand above a dense layer of small seedlings. The stage represented by group 4 differs from the others in the low number of seedlings even though the trees are large. In this group the relative dominance of *Prunus serotina* is low and the canopy is formed mainly by other species, such as *Fagus grandifolia, Acer saccharum* or *Tsuga canadensis* as can be seen in the phytosociological relevés.

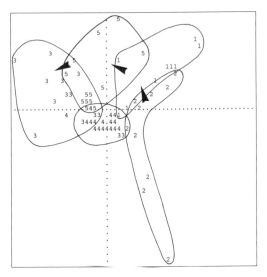

Fig. 3:   Scatter diagram of the PCA of the demographic records. Records from NW-Pennsylvania. First and Second Axis. Numbers indicate groups found in the cluster analysis. Each number stands for one or more records.

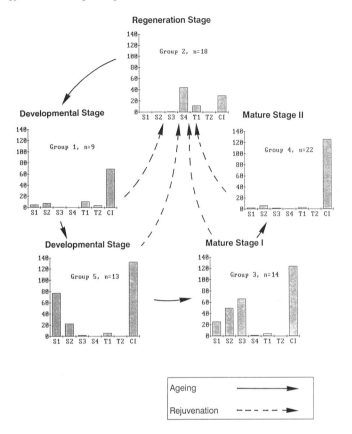

Fig. 4: Succession model for the Prunus serotina population in the Allegheny National Forest, NW-Pennsylvania. Stages are groups found in the cluster analysis.· S1 - S4: Seedling classes (numbers per subplot of 15 m²), T1 - T2: Trees and saplings, living resp. dead (numbers per plot of 225m²), CI - mean circumference at breast height of T1 and T2 (cm)

b) Berlin

Fig. 5 shows the mean numbers of each demographic unit per record. There is a decline from one size class to the next except in the case of the second class. This can be understood as a consequence of the time necessary to grow up to the next size class, which is shorter for S2 than for S3 and S4. Compared to the data from Pennsylvania, the numbers of most demographic units are higher, but the mean size of the trees is smaller. The distribution of the size classes differs between the records in the same way as in the American data set.

The cluster analysis of the demographic units (fig. 6) shows a similar structure to the one of the American data set: The small seedlings (S1 - S3) are grouped closely together with the girth of the trees (CI) but separate from the larger seedlings (S4) and from the number of trees (T1, T2).

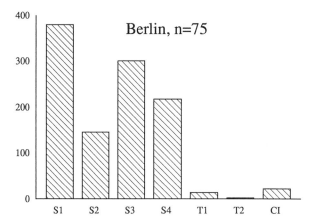

Fig. 5:   Mean numbers of demographic units per sample plot (225 m²). Records from Berlin. S1 - S4: Seedling classes (numbers per subplot of 15 m²), T1 - T2: Trees and saplings, living resp. dead (numbers per plot of 225m²), CI - mean circumference at breast height of T1 and T2 (cm)

Cluster analysis and PCA of the records lead to the ordination diagram (fig. 7) and the succession scheme (fig. 8) in the same way as with the data set from Pennsylvania. The population development runs from a phase with high numbers of large seedlings (gr. 4) over a phase with many small trees (gr. 3) to a mature stage with few relatively large trees and high numbers of small seedlings (gr. 5). One group of records (gr. 2) that also contains relatively large trees has only low numbers of small seedlings; in this group the relative abundance of *Prunus serotina* is low and other species, mainly *Quercus robur, Q. petraea* and *Pinus sylvestris* dominate the canopy.

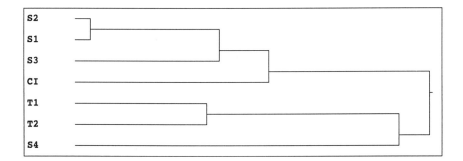

Fig. 6:   Cluster Analysis of demographic units. Records from Berlin. Resemblance measure: correlation coefficient, min. var. clustering.

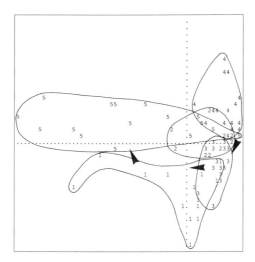

Fig. 7: Scatter diagram of the PCA of the demographic records. Records from Berlin. First and Second Axis. Numbers indicate groups found in the cluster analysis. Each number stands for one or more records.

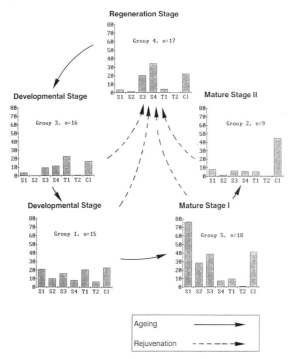

Fig. 8: Succession model for the Prunus serotina population in Berlin. Stages are groups found in the cluster analysis. S1 - S4: Seedling classes (numbers per subplot of 15 m²), T1 - T2: Trees and saplings, living resp. dead (numbers per plot of 225m²), CI - mean circumference at breast height of T1 and T2 (cm)

Discussion

The cluster analysis of size classes showed the spatio-temporal distribution of the small seedlings to be different from the larger ones. This indicates a behaviour known for a number of species of forest trees (SILVERTOWN 1987): having short-lived seed these trees do not form seed banks in the ground like many other plants do, but instead build up a layer of regeneration underneath the canopy that can be called seedling banks. These seedlings can survive for a long time under adverse conditions (i. e. poor light) without height growth; only when the light conditions change, like after windfall or clear-cutting of the canopy, will these seedlings start to grow up to eventually form the next generation of canopy trees. This behaviour has been termed "Oskar Syndrome" by SILVERTOWN (1987) after the character in the novel "Die Blechtrommel" by G. Grass, who stops growing at the age of three. The advantage of such an "Oskar strategy" lies in the fact that a tree species does not need to conquer a site after it becomes favourable for tree growth, but instead preempts the space and can thus be more successful than its competitors. So one might expect *Prunus serotina* to be a very successful species of old-growth forests. Its requirement for high light intensities in the sapling stage, however, limits its success in ecosystems that contain shade tolerant tree species such as beech or some maples. Consequently, *Prunus serotina* reaches dominance in the canopy only for a limited time during succession (HOUGH 1960). In old-growth forests in eastern North America it occurs with low frequencies only (LUTZ 1930, HOUGH & FORBES 1943).

This ecological behaviour is also found in the demographic analysis described here. The trend of the population towards a stage with few large trees can be used to predict possible future population developments. In the case of the data from Pennsylvania, it can be predicted that *Prunus serotina* will remain in the vegetation, but without further disturbance its frequency will become lower. Several authors have shown that *Prunus serotina* is most often favoured by disturbance of the ecosystem - be it man-made or natural (e.g. MCBRIDE 1973, MACKEY & SIVEC 1973, SKEEN 1976).

In southern Wisconsin, *Prunus serotina* invaded oak forests that originated from oak savannahs after white settlers were able to control the formerly frequent fires. In this new ecosystem *Prunus serotina* is a frequent constituent of the understorey. It was the object of a series of studies (AUCLAIR & COTTAM 1971, 1973, MCCUNE & COTTAM 1985). The authors predict that it will remain an important member of these forests, but find it difficult to predict its exact future role in the vegetation.

A comparison of the population biology of *Prunus serotina* in Pennsylvania and in Berlin is difficult because the ecological conditions and the vegetation of the two ecosystems are very different. The role of *Prunus serotina* as a successional species in Pennsylvania is due to the fact that succession leads to dominance of shade-tolerant trees. But this is not the case in Berlin. Here the annual precipitation does not allow shade-tolerant trees like the European beech to become dominant. Oaks and pine, that build up the canopy, cannot regenerate successfully in the shade.

In spite of these differences the two demographic analyses show that the ecological behaviour of *Prunus serotina* is principally the same in both ecosystems. The cluster analysis of the demographic data from Berlin indicates the Oskar strategy and the succession diagram shows the same trend towards a situation with few large individuals and little regeneration like the population in Pennsylvania. The large seedlings (class S4), however, are present in all stages of the population development in Berlin. This can be explained by better growth conditions under a light canopy of oak and pine: whereas the "Oskars" remain in size class S3 in Pennsylvania they grow up to S4 in Berlin.

The population model derived from the PCA of the Pennsylvanian data illustrates how *Prunus serotina* decreases in frequency and dominance in the course of succession. The basic similarity of the two models indicates that the same trend is present in the population in Berlin. So it can be concluded from the comparison that frequency and dominance of *Prunus serotina* in the forests of Berlin will decrease in the course of time, but that the species will remain in the vegetation.

A similar succession model for a vegetation dominated by one tree species was developed by PETERS & OHKUBO (1990) for a Japanese beech forest. Even though *Fagus japonica* is an extreme K-strategist and *Prunus serotina* is more a pioneer species, the two models both stress the importance of the disturbance for the vegetation succession.

The native members of the oak-pine plant community do not show Oskar behaviour. One exception are the maple species (*Acer pseudoplatanus* and *A. platanoides*) that were originally rare or absent and are only recently spreading (SACHSE 1989). So the Oskar strategy seems to be a determinant of the success of *Prunus serotina*. It is supplemented by other population biological characteristics described in more detail elsewhere (STARFINGER 1990). They include its ability to germinate under a wide range of ecological conditions, an early onset of flower and seed production and the transportation of seeds over large distances through bird dispersal.

In conclusion it can be said that the population biological characteristics of *Prunus serotina* have made its successful invasion to Central Europe possible. The species will maintain a population in the future. During limited phases of its population development it can be abundant and may have to be controlled for nature conservation and forestry purposes. In the long run it will not be a major constituent of the canopy but will decrease in frequency and abundance.

Zusammenfassung

Die Spätblühende Traubenkirsche (*Prunus serotina*), eine aus Nordamerika stammende Waldbaumart, hat sich in Mitteleuropa erfolgreich eingebürgert. Indem sie dichte Strauchschichten in vorher lichten Wäldern aufbaut, kann sie einheimische Pflanzenarten verdrängen und damit zu einem Problem für den Naturschutz werden.

In ihrer ursprünglichen Heimat, in NW-Pennsylvania, und in Berlin wurden demographische Aufnahmen angefertigt und mit Hilfe multivariater Methoden analysiert. In beiden Gebieten weist die zeitlich-räumliche Verteilung der Größenstadien auf ein sog. Oskar-Verhalten der Traubenkirsche hin: die Jungpflanzen können im Schatten der Mutterbäume mit begrenztem Wachstum lange Zeit überleben. Diese Verhalten ist - zusammen mit anderen populationsbiologischen Merkmalen - eine wichtige Voraussetzung für die erfolgreiche Besiedlung neuer Wuchsorte und für die Einbürgerung in Mitteleuropa.

Die Entwicklung der Population verläuft in beiden Untersuchungsgebieten über unterschiedliche Phasen. Ohne Störungseinfluß führt die Sukzession zu einem Stadien mit wenigen großen Traubenkirschen in der Baumschicht und mit wenig Verjüngung. Dieser Trend ist trotz sehr unterschiedlicher Vegetation in Berlin ebenso wie in Pennsylvania aus der demographischen Analyse abzulesen. Daraus wird die Prognose abgeleitet, daß *Prunus serotina* in Berlin erhalten bleiben wird, aber geringere Frequenz haben wird als zur Zeit.

References

ANONYMUS (1982): Forstliche Rahmenplanung (Materialien). Herausgegeben v. d. Berliner Forsten, Landesforstamt.

AUCLAIR, A. N. & COTTAM, G. (1971): Dynamics of Black Cherry (*Prunus serotina* Ehrh.) in Southern Wisconsin oak forests. Ecological Monogr. 41, 153-177.

AUCLAIR, A. N. & COTTAM, G. (1973): Multivariate analysis of radial growth of Black Cherry (*Prunus serotina* Ehrh.) in Southern Wisconsin oak forests. Am. Midl. Natur. 89, 408-425.

BRAUN-BLANQUET, J. (1964): Pflanzensoziologie. 3. Aufl. Wien & New York.

COLLINS, B. S. & PICKETT, S. T. A. (1982): Vegetation composition and relation to environment in an Allegheny hardwood forest. Am. Midl. Natur. 108, 117-123.

DRAKE, J. A. *et al.* (eds.) (1989): Biological invasions. A global perspective. J. Wiley & Sons, Chichester.

EIJSACKERS, H. & OLDENKAMP, L. (1976): Amerikaanse vogelkers, aanwaarding of beperking? Landbouwkundig Tijdschr. 88, 366-375.

ELTON, C. S. (1958): The ecology of invasions by animals and plants. London.

ERNST, K. (1965): Späte Traubenkirsche und Traubenholunder. Berliner Natursch. Bl. 26, 4-11.

FUKAREK, F. (1987): Pflanzen in Ausbreitung. Gefährdete Arten - gefährliche Arten. Bot. Rundbr. Bez. Neubrandenburg 19, 3-8.

GROVES, R. H. & BURDON, J. J. (1986): Ecology of biological invasions. Austr. Acad. of Science, Canberra.

HARPER, J. L. (1977): Population biology of plants. London.

HOUGH, A. F. & FORBES, R. D. (1943): The ecology and silvics of forests in the high plateaus of Pennsylvania. Ecol. Monogr. 13, 299-320.

HOUGH, A. F. (1960): Silvical characteristics of Black Cherry, *Prunus serotina*. Station Paper No 139, NE Forest Exp. Sta. USDA Forest Service, Upper Darby, PA.

JÄGER, E. (1988): Möglichkeiten der Prognose synanthroper Pflanzenausbreitungen. Flora 180, 101-131.

JOENJE, W., BAKKER, K. & VLIJM, L. (eds.) (1987): The ecology of biological invasions. Proc. K. Ned. Akad. Wet. Ser. C 90, 1-80.

KORNBERG, H. & WILLIAMSON, M. H. (1986): Qunatitative aspects of the ecology of biological invasions. Phil. Trans. R. Soc. London B 314.

KRAUSS, M., LOIDL, H., MACHATZI, B. & WALLACHER, J. (1990): Vom Kulturwald zum Naturwald. Landschaftspflegekonzept Grunewald. Veröffentlichungsreihe Berliner Forsten Bd. 1.

LUTZ, H. J. (1930): The vegetation of Heart's Content, a virgin forest in northwestern Pennsylvania. Ecology 11, 1-29.

MACKEY, H. E. & Sivec, N. (1973): The present composition of a former oak-chestnut forest in the Allegheny mountains of western Pennsylvania. Ecology 54, 915-919.

MCBRIDE, J. (1973): Natural replacement of disease killed elms. Amer. Midl. Natur. 90, 300-306.

MCCUNE, B. & COTTAM, G. (1985): The successional status of a southern Wisconsin oak woods. Ecology 66, 1270-1278.

MOONEY, H. A. & DRAKE, J. A. (eds.) (1986): Ecology of biological invasions of North America and Hawaii. Springer New York.

MOONEY, H. A. & DRAKE, J. A. (1989): Biological invasions: a SCOPE program overview. In DRAKE, J. A. *et al.* (eds.), Biological invasions. A global perspective. J.Wiley & Sons, Chichester, New York, Brisbane, Toronto & Singapore.

PETERS, R. & OHKUBO, T. (1990): Architecture and development in *Fagus japonica - Fagus crenata* forest near Mount Takahara, Japan. Journal of Vegetation Science 1, 499-506.

SACHSE, U. (1989): Die anthropogene Ausbreitung von Berg- und Spitzahorn. Ökologische Voraussetzungen am Beispiel Berlins. Landschaftsentwicklung u. Umweltforschung 63, 1-132. TU Berlin.

SILVERTOWN, J. W. (1987): Introduction to plant population ecology. 2nd ed. London.

SKEEN, J. N. (1976): Regeneration and survival of woody species in a naturally created forest opening. Bull. Torrey Bot. Club 103, 259-265.

STARFINGER, U. (1987): Bericht zum Forschungsvorhaben: Die Spätblühende Traubenkirsche (*Prunus serotina* Ehrh.) in Berlin (West). Manuskript. Berlin n.p.

STARFINGER, U. (1990): Die Einbürgerung der nordamerikanischen Spätblühenden Traubenkirsche (*Prunus serotina* Ehrh.) in Mitteleuropa. Landschaftsentwicklung u. Umweltforschung 69, 1-136. TU Berlin.

SUKOPP, H., BLUME, H.-P., ELVERS, H. & HORBERT, M. (1980): Beiträge zur Stadtökologie von Berlin (West). Exkursionsführer für das zweite europäische ökologische Symposium. Landschaftsent. u. Umweltforsch. 3, 1-225.

TAYLOR, N. (1928): The vegetation of the Allegany State Park. NY State Museum Handbook No 5.

TWEEL, P. A. van den & EIJSACKERS, H. (1987): Black Cherry, a pioneer species or 'forest pest'. Proc. K. Ned. Akad. Wet. Ser. C 90, 59-66.

WEIN, K. (1930): Die erste Einführung nordamerikanischer Gehölze in Europa. Mitt. Dtsch. Dendrol. Ges. 42, 137-163.

WILDI, O. & ORLOCI, L. (1983): Management and multivariate analysis of vegetation data. Ber. Eidgen. Anst. forstl. Versuchswesen Nr. 215.

WILDI, O. & ORLOCI, L. (1988): Mulva 4, a package for multivariate analysis of vegetation data. Mskr.

Species Conservation: A Population-Biological Approach
A. SEITZ & V. LOESCHCKE (eds.) © 1991 Birkhäuser Verlag, Basel

An Experiment on Dynamic Conservation of Genetic Resources With Metapopulations

J.P. Henry [1], **C. Pontis** [2], **J. David** [2] and **P.H. Gouyon** [3]

[1] Institut National Agronomique Paris-Grignon. Chaire de Génétique Évolutive et Amélioration des Plantes. 16, rue Claude-Bernard, 75005 Paris, France.
[2] Station de Génétique Végétale INRA-UPS, Ferme du Moulon, 91190 Gif-sur-Yvette, France
[3] Université de Paris-Sud. Laboratoire d'Évolution et Systématique des Végétaux. Bât. 362., 91400 Orsay, France.

Abstract

From a conservation perspective, the genetic variability of natural populations that must be preserved is mainly that of selected genes, which are involved in adaptation and fitness. It is suggested that guidelines given by theoretical models suited for neutral genetic variability are inadequate for selected genes, and that Wright's "Shifting Balance Theory" is more relevant. Therefore, a metapopulation structure which permits a certain amount of local differentiation is suggested.

Cultivated plants are a good model to test different management methods to preserve genetic diversity. An experiment on dynamic conservation of genetic resources of wheat with a large genetic base is being grown in a multi-site experimental network. In 1984, each initial population was issued to 7 to 12 locations, where they are grown each year under two farming conditions. Each year, a part of the harvest is sown again in the same location and condition, and the rest is conserved for future analysis. Genetic variability of these evolving populations is investigated for seed protein genes, powdery mildew resistance genes. dwarfing genes, and for several quantitative traits that are selectively important. First results show a significant divergence of populations after four years. Some alleles have been lost in some populations, yet genetic diversity is preserved on the whole. A very highly significant excess of heterozygote plants is found for several genes. Possible causes of these observations are discussed.

In this experiment, an extinction-recolonization process should take place. Alternative ways to perform this are considered.

Introduction

We think that experiments in dynamic conservation of genetic resources of cultivated plants could provide useful guidelines for the management of wild populations. The following features make cultivated plants a good model for researching the best ways to manage populations in order to maintain genetic variability:

-   In several of these species, genetic variability has been thoroughly studied (e.g. HELENTJARIS 1987 for maize, RICK & YODER 1988 for tomato, WORLAND *et al.* 1987 for wheat).

-   They can be easily handled and it is possible to constitute artificial synthetic populations with known genetic diversity.

-   Loss of variability in these synthetic populations is not as damaging as it would be for endangered species, thus different ways of management can be tested.

Much work has already been done concerning the conservation of genetic resources of cultivated species and their wild relatives (FORD-LLOYD & JACKSON 1986, BROWN *et al.* 1989). On a world scale, the most used method is "static" conservation in seed banks (but there are very few in France). It is probably the only practicable way from a short term perspective and necessary as an emergency measure, and it is often viewed as a security measure. But it has serious drawbacks: seed banks are costly (I.B.P.G.R. 1985), prone to disasters, difficult to manage (PERNES 1984); longevity of seeds is not infinite, very short for some species (ROBERTS 1975). Sooner or later, it will be necessary to regenerate sets of seeds by sowing, and during this stage, the genetic constitution of heterogeneous populations will be affected by natural selection and genetic drift in an unpredictable way. Therefore, it is to be feared that static conservation will be inefficient in the long run, since in such a way, genetic variability can only be lost. Even if we could overcome these difficulties, we think, on the grounds of the "Red Queen hypothesis" (VAN VALEN 1973), that, if genetic variability is "frozen" too long in seed banks, some risks exist that material entered into the banks will become maladapted to the future biotic environment (competitors, parasites, predators...) which continues to evolve.

On the other hand, a large amount of genetic variability exists in natural populations (especially revealed by electrophoretic studies, reviewed for plants by BROWN 1979, HAMRICK 1983, LOVELESS & HAMRICK 1984): without human intervention , nature not only keeps genetic resources, but creates new ones. Of course, in natural populations, at each time some genotypes are lost, but others are created by mutation and recombination, or introduced by migration from other populations. It is a dynamic system, so this perpetual evolution allows populations to remain adapted to their environment.

The idea of "dynamic" conservation of genetic resources is therefore very simple: we would like to do the same as nature does without us! Surprisingly, this is quite difficult, both for theoretical and practical reasons and actually, dynamic conservation is rather a complement to than a substitute for static conservation.

Compared with static conservation, relatively little research has been done in dynamic conservation of genetic diversity of cultivated plants. The most impressive experiment in this field was done on synthetic populations of barley cultivated over a period of more than 40 years. An invaluable amount of information about the evolution of diversity has been gathered (KAHLER et al. 1975, DE SMET et al. 1985, briefly reviewed in BROWN 1983).

On the other hand, a number of theoretical models have been built to predict evolution of genetic diversity in populations with known demographic and genetic features. Especially relevant to a conservation perspective are those referring to small populations (e.g. ALLENDORF 1983, VARVIO et al. 1986, MARUYAMA & FUERST 1985a,b), and guidelines have been suggested for management of such small populations. Most of these are based on the concept of effective size (WRIGHT 1931, 1969 chapter 8), which, according to these guidelines, should be kept as great as possible in order to minimize inbreeding depression and genetic drift by which genetic diversity is lost; demographic bottlenecks should be particularly avoided because they have devastating effects on the amount of genetic diversity in populations.

This is unquestionably true for selectively neutral genes: for these genes, the level of variability in a population results only from an equilibrium between mutation and migration from other populations, which both increase allelic diversity, and genetic drift which reduces it. But in conservation, we are mainly interested in selected genes, which are implicated in adaptation and fitness. For them, and especially for groups of interacting genes, things are far less clear.

One of the most relevant theories for them, we think, is the "Shifting Balance Theory" of WRIGHT (1931, 1982). It may throw a different light on the problem of conservation of genetic diversity. This theory was coined through the now-famous "adaptive landscape" metaphor: in the (multi dimensional) space of genotypes, where the orthogonal axes (abscissa) are the alleles frequencies at each locus (this was not very clear in the early writing of WRIGHT see CROW 1987 for a discussion), we can erect at each point an additional dimension (ordinate) proportional to the mean fitness of a randomly mating population with the combination of allele frequencies at this point. We thus get a fitness surface or "adaptive landscape" with peaks where allelic combinations give local maximal fitness, separated by maladaptive valleys where there is minimal fitness. The existence of multiple maxima results from non-additive interactions between genes with respect to fitness, which are thought to be very frequent (WRIGHT 1982). From an initial genetic composition, natural selection pushes the population to the nearest peak on the adaptive landscape. In fact, populations are held captive on the peaks by natural selection, like air-bubbles in a spirit-level. How could a population evolve from one good gene combination (a peak) to a better one, via a maladaptive valley? For this to occur, a population needs some random effect due to genetic drift, for example during a demographic bottleneck. So, in the wrightian view, bottlenecks and drift may create genetic novelty and be good for species. This is the exact opposite of what the theory for conservation of selectively neutral genes advocates!

We do not suggest, here, that deliberate bottlenecks should be made in endangered species, nor do we hint that demographic bottlenecks are not very risky for the survival of a single population. Obviously they are, both for demographic reasons (stochastic risks of extinction) and genetic ones (inbreeding depression). We just want to emphasize that it is not advisable to rely uniquely on theories suited to neutral genetic variability to preserve selected genetic variability. Rules of management can be different for different types of genetic diversity: selected vs. neutral, quantitative vs. single-locus... This idea has already been emphasized by LANDE & BARROUCLOUGH 1987, but in a different way and with different conclusions.

Moreover, demographic bottlenecks are usual events for populations of many species of plants and animals, especially those living in disturbed environments. Extinction of local populations often results from them. However, it is temporary because migrants from other populations can come into the site or into newly created ones and found new populations (for examples, MERRIAM 1988, HARRISON *et al.* 1988). Therefore extinction of local populations is something very different from the extinction of the whole species, at least as long as man has not built impassable barriers between populations ; if so, we will see extinction of each local population, one after the other and without recolonization, until the extinction of the whole species.

Finally, we suggest three proposals from a genetic point of view:

1. Dynamic conservation of significant adaptive genetic variability would probably be more effective when using a population structure which permits a certain amount of local genetic differentiation. If there are many differentiated subpopulations, some of them could randomly drift towards new favorable adaptive combinations. Then, the growing subpopulation concerned would be able to export migrants and contribute to an upgrading of the whole (according to WRIGHT, quoted in CROW 1987).

2. We must not think of conservation at the level of a single population, but at that of sets of local populations able to exchange some migrants including the extinction-recolonization process, that is to say "metapopulations" (LEVINS 1970).

3. It is important to preserve the possibility of migration between populations, or alternatively to restore it if it is lost.

These proposals are not new: proposals 2 and 3 are akin to those made by other conservation biologists, usually more on demographic and ecological grounds (see GILPIN 1987). Proposal 1 is more controversial (see HANSKI 1989 for a discussion from a demographic point of view and references therein).

The following experiment has been designed on the basis of proposal 1.

Presentation of the Experiment (Material and Methods)

Three synthetic populations of wheat are grown in a multi-site experimental network. This experiment associates 5 agronomical universities, 11 agricultural secondary schools and the French National Institute for Agronomical Research (I.N.R.A.), with the financial support of the French Ministry for Agriculture, Water and Forestry. One important aim of this experiment is educational training of teachers and students in population biology, and growing public awareness of the problem of genetic resources in agricultural management circles.

Plant Material

Wheat (*Triticum aestivum* L.) has been chosen for this experiment because the selfing of this crop makes less stringent requirements for spatial isolation of populations in farm-schools, and primarily because synthetic populations were already available at the time when we had the opportunity to start this experiment (1984). We have used three composite populations made for other purposes by Doussinault & Trottet (INRA station in Rennes), Rousset & Pichon (INRA station in Clermont-Ferrand) and Picard (INRA station in Gif), from pyramidal crosses of many parents: population PA was made by crossing 16 parental lines (ROUSSET *et al.* 1988), PB with 16 other parental lines of more "exotic" origin than PA (PICARD *et al.* 1988), and PS with 55 ones (TROTTET 1988). Moreover population PS contains the recessive male-sterility nuclear gene *ms1b* (MCINTOSH 1988) from the variety "Probus" (FOSSATI & INGOLD 1970). Concerning this latter population, only grains from male-sterile (female) plants are sown each year, so a 50% male-sterile (*ms1b/ms1b*) - 50% male-fertile (*Ms1b/ms1b*) disjunction is preserved in each generation. The loss of the male-sterility gene *ms1b* is thus prevented, and this population is partly allogamous. Note that, unfortunately, possible differences in the evolution of variability in autogamous (PA, PB) and allogamous (PS) populations will be difficult to attribute to an effect of the mating system, owing to differences in numbers of parents.

Experimental Design

These three populations with large genetic bases (referred to as "pools" from now on) were dealt out in 1984 among agricultural schools to be grown in their associated farms, the responsibility and control lying with volunteer teachers collaborating in this experiment. Each population is grown using two farming methods:

-     "intensive", which is similar to the method usually applied for wheat on these farms ; however sowing density is low (150 grains/m²) to prevent the competition from becoming too unfair to genotypes with low height ;

-    "extensive", with the same density, one third of the nitrate fertilizers used in the intensive method, no fungicide.

Each of these populations is cultivated isolated from any other wheat cultures by at least 30m for PA and PB, and 150m for PS. For each, a 100 m² area was recommended to obtain around 10000-15000 plants by isolated essay, but, due to a variety of accidents, the number of effectively grown and harvested plants has often been much lower, sometimes no more than a few hundred, and in two cases it was zero, that is to say extinction (see the following discussion).

Locations are well distributed throughout France (see map Fig.1), although, as some farms could not cultivate six populations, each pool is not represented in each locality (PA: 7 locations, PB:9 ,PS:12). Seeds are harvested separately every year for each location and each cultivation method. Part of the harvest is sown the following autumn on the same farm and under the same conditions, and the rest is conserved in a cold room for future analysis. For PS populations, male-sterile plants are identified and marked at the time of flowering ; the person in charge inspects the population every 2 or 3 days during the time of flowering, to avoid selection of a precocious class of male-sterile plants. Around 2000 male-sterile

Fig. 1:   Localities of the experimental network where the wheat populations are grown. Comparative studies for agronomical traits take place in "Le Moulon".

plants are marked and will be separately harvested for the next sowing, as mentioned previously. We aim to continue this experiment as long as possible.

Analysis of Genetic Variability

Two types of genetic variability are being investigated in the conserved part of the harvests: electrophoretic single-locus variability, which is presumably more or less neutral (KIMURA 1983); variability for agronomically and selectively important traits which is mainly quantitative, such as height, precocity, grain production. In addition, diversity for known selected loci is studied: genes for powdery mildew resistance, and dwarfing genes (*Rht1* and *Rht2*: GALE & YOUSSEFIAN 1985) which are present in some parental lines.

Regarding electrophoretic variability, preliminary studies have been performed among parental lines of the composite crosses to investigate the potential diversity likely to be present in populations. Firstly, enzymatic systems have been studied (LACAZE 1986). Because of quite low diversity disclosed by classical electrophoresis, as is usual for wheat (GALE & SHARP 1988), composite populations have still not been analyzed for these systems. Secondly, diversity of parental lines for seed proteins (glutenins and gliadins) has been studied by SDS PAGE electrophoresis (methods in BRANLARD *et al.* 1990 for gliadins, PAYNE *et al.* 1979 for glutenins). The genetic basis of glutenin and gliadin diversity in wheat is known (reviewed in MCINTOSH 1988). These analyses have shown a large amount of genetic variability (see results) so that populations themselves are being analyzed, for two generations: the initial pools which had been dealt out among localities (subsequently referred to as PA_0, PB_0 and PS_0), then every population cultivated in each location and cultural condition, four generations after the initial deal (30 individuals for each in a first step, and then as many as 70).

For quantitative traits, grains from 3900 spikes selfed in 1988 in the different locations were sown in 2 lines by spike in a randomized experimental plot in the INRA station "Le Moulon" (Gif-sur-Yvette). In 1989, data was collected in these lines for some quantitative agronomical traits: precocity of spike emergence, lodging, height, spike features (length, number of spikelets, number of grains), grain weight, humidity and specific density of grains. In addition, the lines were marked for possible intra-line disjunctions of bearded vs. unbearded spike traits, due to possible heterozygosity of the parental plant for this trait. Lineages from each initial selfed spike were studied again in 1990, in 3 m² experimental plots, for the same quantitative traits as 1989 and for yield. Analysis of variance (ANOVA) was performed to find out the part of inter- and intra-population diversity for the various studied traits, and the importance of geographical localization and cultivation conditions in their evolution. For allogamous PS populations, homozygous lines were extracted by "single seed descent" (SSD), and will be studied later in the same way.

First Results and Discussion

We present here only some preliminary undetailed results from a first analysis (DAVID and PONTIS unpublished); these are still far from complete.

Large allelic diversity exists for glutenin and gliadin genes in each pool (PA, PB and PS): 3 glutenin loci were found: *GluA1* with 3 alleles, *GluB1* with 4 or 5 alleles depending on pool, *GluD1* with 3 or 4 alleles. Two gliadin loci were recognized: *GliB1* with only 2 distinguishable alleles, and *GliD1* with 4 to 7 alleles .

Tab. 1.: Results of electrophoretic analysis for the GliD1 locus in populations of the PB pool. (Electrophoretic data by C. PONTIS, unpublished). The eight alleles have been pooled in two allelic groups (1 and 2), because band superpositions may confuse some genotypes inside each allelic group, but not between them. Populations are specified by their location (see Fig. 1) and cultural conditions (E: extensive; I: intensive). Locations have been ordered according to their mean allelic frequency.

N: analyzed sample size; n11: number of homozygous plants for allele 1; n12: number of heterozygous plants 12; n22: number of homozygous plants for allele 2; p: allele 1 frequency; SE: standard-error on $p = \sqrt{p(1-p)(1+F)/(2N-1)}$; Htz: percentage of heterozygous plants; F: intra-population fixation index ; F_{is}, F_{st}, F_{it}: Wright's F statistics for this set of samples (excluding PB_0), calculated using Nei's formulas: NEI 1987 pp.161-162 (F_{is}: mean intra-population fixation index; F_{st}: between-population differentiation; F_{it}: fixation index on the whole). F_{is}', F_{st}', F_{it}': Estimations of Wright's F statistics for the PB pool, assuming equal size for every local population (using Nei's formulas: NEI 1987 pp.164-165).

Population		N	n11	n12	n22	p ± SE	Htz(%)	F
PB_0		69	28	13	28	0.50 ± 0.05	18.84	0.62
(starting population)								
Nérac	E	51	13	0	38	0.25 ± 0.06	0.00	1.00
Nérac	I	49	10	3	36	0.23 ± 0.06	6.12	0.83
Toulouse	E	32	7	2	23	0.25 ± 0.07	6.25	0.83
Toulouse	I	64	23	3	38	0.38 ± 0.06	4.69	0.90
Montreuil	E	29	10	1	18	0.36 ± 0.09	3.45	0.93
Le Chesnoy	E	68	30	3	35	0.46 ± 0.06	4.41	0.91
Le Chesnoy	I	73	30	2	41	0.42 ± 0.06	2.74	0.94
Rennes	E	43	23	6	14	0.60 ± 0.07	13.95	0.71
Rennes	I	52	24	0	28	0.46 ± 0.07	0.00	1.00
Le Moulon	E	68	37	0	31	0.54 ± 0.06	0.00	1.00
Le Moulon	I	66	34	1	31	0.52 ± 0.06	1.52	0.97
Venours	E	62	30	3	29	0.51 ± 0.06	4.84	0.90
Venours	I	58	30	10	18	0.60 ± 0.06	17.24	0.64
Châlons	E	64	49	0	15	0.77 ± 0.05	0.00	1.00
Châlons	I	75	50	5	20	0.70 ± 0.05	6.67	0.84
Total		854	400	39	415	0.49		4.57	0.91
(excluding PB_0)								

$F_{is} = 0.90$ $F_{st} = 0.09$ $F_{it} = 0.91$

$F_{is}' = 0.92$ $F_{st}' = 0.08$ $F_{it}' = 0.93$

The initial variability present in the composite pools, as revealed by electrophoretic analysis of parental lines and initial generation, has been on the whole preserved until now: no allele has been completely lost, even among the rarest ones, excepting one allele of *GliD1* in the PB pool, found in one plant among 70 in PB_0 and not found in subsequent populations. Yet, frequently, some alleles were lost in some local populations, as far as can be stated with the limited number of analyzed plants (with 30 plants, assuming complete homozygosity for advanced generations of an autogamous population, the probability for no detection of an allele is greater than 5% when the frequency of this allele is smaller than 9.5%)

Some evolution of genetic constitution has taken place which makes populations significantly divergent, both in regard to electrophoretic variability and many quantitative traits. Data on the *GliD1* locus in the PB pool are presented as an example in Tab. 1. In this case the mean allelic frequencies, calculated by pooling all populations descended from PB, have remained close to the initial ones in PB_0 (for some other loci, shifts of mean allelic frequencies are observed). This example is peculiar in that allelic frequencies exhibit close similarities between the two conditions in the same location, associated with significant differences between some locations. For other electrophoretic loci, and for this locus in other pools, differences between conditions in allelic frequencies are about as large as between locations. Moreover allelic frequencies of this locus in this pool correlate with the latitude of the stations (compare with the map; Spearman's rank correlation coefficient rs = 0.74, significant: $p < 0.05$). Again this peculiarity has not been observed for other grain-protein loci, but geographical structure has been found for some agronomical traits. The special features of this locus suggest that selection (climatic conditions?) acts upon this locus, or upon other loci in linkage disequilibrium with it in this pool.

Interpopulation divergence is given by Wright's statistics F_{st}=0.09. Globally for 4 protein-grain loci, the part of the total genetic diversity which is due to the interpopulation divergence, as given by G_{st} (NEI 1973), is only between 4% to 6% depending on the pools. Similarly, for quantitative traits, interpopulation diversity is quite low, albeit insignificant, as shown by ANOVA: in PB pool, interpopulation level explains only around 5% of the global quantitative variability. This low amount of interpopulation differentiation should be explained keeping in mind that only four years of separate cultivation in different localities have taken place. Owing to this short time and the sizes of populations, we think that this divergence is better explained by different selective pressures in different localities and cultivation conditions, favoring different genotypes, than by genetic drift or foundation effect. Note that genetic contamination by pollen from other fields or by alien seeds in harvesting or sowing operations cannot be completely ruled out, despite spatial isolation of the assays, predominant selfing for PA and PB, and special precautions in harvesting and handling of seeds. However, the absence of alien alleles not already present in the initial cross makes this hypothesis less probable; moreover the alleles which have increased in frequency in some populations are not those present in wheat varieties grown and analyzed in the last years.

Heterozygosity was expected to be very low in advanced generations of autogamous PB populations, since PB_0 when dealt out was already three generations old from the end of the pyramidal cross. Analysis of this PB_0 generation has shown a very highly significant excess of heterozygous plants for all studied loci with respect to expected proportions, knowing parental lines and assuming equal contribution of each line. As we know that breeders had made some selection in initial crosses and the subsequent generation, but not what exactly happened, explanations for this excess can be only very tentative. On the other hand, comparisons can be made between the initial generation PB_0 and descended populations 4 years later, where no conscious selection happened. Assuming strict selfing and no selection, residual heterozygosity after 4 generations should be 1/16 of that present in PB_0. As shown in Tab. 1, results vary with populations, but, on the whole over all pooled populations, a very highly significant excess of heterozygous plants is again observed (χ^2 = 29.0 with 1 df. ; $p < 10^{-3}$), and similar results occur for some other loci and populations. To explain this fact, the hypothesis of strict selfing can be questioned: a low rate of outbreeding is known to occur in wheat as in most other self-fertilizing species. We have calculated that the mean rate of outbreeding should be t = 3.9% to fit the observed decrease of mean heterozygosity throughout the 4 generations. Although such a level cannot be ruled out, especially in the most southern part of France, it is unlikely as a mean rate for wheat (note that even if the two most heterozygous populations, "Venours I" and "Rennes E", are discarded, the excess remains very highly significant: χ^2 = 10.0, $p \approx 10^{-3}$, and the mean outbreeding rate should be t = 3.2%). Therefore, we think that some selection favoring heterozygous genotypes must be invoked. Such results in selfing species have been obtained by many workers (for reviews and discussions, see ALLARD et al. 1968 and BROWN 1979). That does not necessarily imply an overdominance for the studied markers, but it can be explained by a selective advantage for plant heterozygosity in other loci or for most heterozygous plants on the whole genome, since in selfing populations, linkage disequilibrium is known to be strong and frequent, and multilocus heterozygosity to be higher than expected from single locus frequencies.

Finally, we would like to discuss the problem of the extinction-recolonization process which takes place in our experiment. As mentioned previously, two local populations have become extinct. For one (Montpellier), we have chosen to recreate a population by mixing, in equal part, the local populations from all the other locations (the other extinct population, in Valence, has been renewed simply by taking a conserved part of a preceding-year harvest from this place). In fact, we feel that we lack the theoretical basis to choose the best way for recreating extinct populations. The problem will be the same when we establish new populations in new locations. We could also mix populations periodically, alternating with steps of local differentiation : if populations are kept isolated too long, probably the largest part of variability will be lost in each (whether global diversity will be maintained on the whole is another question, which we hope to answer in some years), and clearly the experiment without exchanges would not be fruitful any longer. On the other hand, we think that some local differentiation should take place, as previously explained. But how much, how long between mixing ? Alternatively, we could consider putting a low level of continuous migration (seed exchanges) between some or all populations. What would be better ? Clearly, we need more theoretical and experimental work to choose the best ways to manage populations.

Zusammenfassung

Vom Standpunkt des Artenschutzes ist die zu bewahrende genetische Variabilität natürlicher Populationen hauptsächlich die von ausgewählten Genen, die etwas mit Anpassung und Fitneß zu tun haben. Es wird darauf hingewiesen, daß von theoretischen Modellen abgeleitete Richtlinien, die für neutrale genetische Variabilität entwickelt wurden, nicht geeignet sind für genetische Variabilität, die der Selektion unterliegt. Vielmehr muß Wright's "Shifting Balance Theorie" angewandt werden. Deshalb wird eine Metapopulationsstruktur empfohlen, die ein gewisses Ausmaß lokaler Differenzierungen erlaubt.

Nutzpflanzen sind gute Modelle, um verschiedene Methoden des Managements zur Bewahrung genetischer Vielfalt zu erproben. Es wird ein Versuch zur dynamischen Konzipierung der genetischen Resourcen von Weizen vorgestellt. Drei künstliche Weizenpopulationen mit einer breiten genetischen Basis werden an verschiedenen Stellen gezüchtet. 1984 wurde die Ursprungspopulation an 7 bis 12 verschiedene Orte verschickt, wo sie über das Jahr unter 2 verschiedenen Zuchtbedingungen gehalten werden. Jedes Jahr wird ein Teil der Ernte am gleichen Ort und unter den gleichen Bedingungen wieder ausgesät, und der Rest wird für spätere Untersuchungen aufbewahrt. Die genetische Variabilität dieser sich evolvierenden Populationen wird bezüglich der Gene untersucht, die den Proteingehalt der Samen, Resistenz gegen Mehltau und Kurzwüchsigkeit beeinflussen, sowie einige quantitative Merkmale, die besonders wichtig sind. Erste Ergebnisse zeigen eine signifikante Auseinanderentwicklung der Populationen nach 4 Jahren. Einige Allele sind in einigen Populationen verlorengegangen, jedoch die genetische Diversität als ganzes wurde erhalten. An einigen Genorten wurde eine hohe Signifikanz an Überschuß von heterozygoten Pflanzen gefunden. Mögliche Ursachen dieser Erscheinungen werden diskutiert.

In diesem Experiment soll ein Prozeß von Auslöschung und Wiederbesiedlung stattfinden. Verschiedene Möglichkeiten, um dies zu bewerkstelligen, werden betrachtet.

Acknowledgements :

We are indebted to the teachers, too numerous to be cited here, who are in charge of the experimental network in schools and universities. Without their help and collaboration, this experiment could not have been set up. We are also very grateful to G. Branlard, G. Doussinault, E. Picard, M. Pichon, M. Rousset and M. Trottet who supplied us with the synthetic wheat populations used in this experiment and helped us with their competence. We wish also to thank the organizers of the Symposium, particularly Dr. Seitz and Dr. Loeschcke. This work is supported by the DGER of "Ministère de l'Agriculture" of France and directed by A. Gallais.

References:

ALLARD, R.W. ; S.K. JAIN & P.L. WORKMAN 1968. The genetics of inbreeding populations. Adv. in Genet. 14:55-131.

ALLENDORF, F.W. 1983. Isolation, gene flow, and genetic differentiation among populations. Pp. 51-65 in: C. M. SCHONEWALD-COX, S.M. CHAMBERS, B. MACBRYDE & W.L. THOMAS (eds.) Genetics and Conservation. A Reference for Managing Wild Animal and Plant Populations. The Benjamin/Cummings Publ. Co., Menlo-Park, Calif.

BRANLARD, G. ; B. PICARD & C. COURVOISIER 1990. Electrophoresis of gliadins in long acrylamide gels : method and nomenclature. Electrophoresis 11:310-314.

BROWN, A.H.D. 1979. Enzyme polymorphism in plant populations. Theor. Pop. Biol. 15:1-42.

BROWN, A.H.D. 1983. Barley. Pp. 55-77 in S.D. TANKSLEY &,T.J. ORTON (eds.) Isozymes in Plant Genetics and Breeding, Part B. Elsevier, Amsterdam.

BROWN, A.H.D. ; O.H. FRANKEL, D.R. MARSHALL & J.T. WILLIAMS (eds.) 1989. The Use of Plant Genetic Resources. Cambridge University Press, Cambridge.

CROW, J.F. 1987. Population genetics history : a personal view. Ann. Rev. Genet. 21:1-22.

DE SMET, G.M.W. ; A.L. SCHAREN & E.A. HOCKETT 1985. Conservation of powdery mildew resistance genes in three composite cross populations of barley. Euphytica 34:265-272.

FORD-LLOYD, B. & M. JACKSON 1986. Plant Genetic Resources : an Introduction to their Conservation and use. Edward Arnold, London.

FOSSATI, A. & M. INGOLD 1970. A male sterile mutant in *Triticum aestivum*. Wheat Info. Serv. 30:8-10.

GALE, M.D. & P.J. SHARP 1988. Genetic markers in wheat - developments and prospects. Pp. 469-475 in: T.E. MILLER & R.M.D. KOEBNER (eds.) Proceedings of the Seventh International Wheat Genetics Symposium held at Cambridge, UK, 13-19 July 1988. Vol. 1. Institute of Plant Science Research. Cambridge

GALE, M.D. & S. YOUSSEFIAN 1985. Dwarfing genes in wheat. Pp. 1-35 in: G.E. RUSSEL (ed.) Progress in Plant Breeding 1. Butterworths, London.

GILPIN, M.E. 1987. Spatial structure and population vulnerability. Pp.125-139 in: SOULÉ, M.E. (ed.) Viable Populations for Conservation. Cambridge University Press, Cambridge.

HAMRICK, J.L. 1983. The distribution of genetic variation within and among natural plant populations. Pp 335-348 in: C. M. SCHONEWALD-COX, S.M. CHAMBERS, B. MACBRYDE & W.L. THOMAS (eds.) Genetics and Conservation. A Reference for Managing Wild Animal and Plant Populations. The Benjamin/Cummings Publ. Co., Menlo-Park, Calif.

HANSKI, I. 1989. Metapopulation dynamics: does it help to have more of the same ? Trends in Ecol. and Evol. 4:113-114.

HARRISON, S.; D.D. MURPHY & P.R. EHRLICH 1988. Distribution of the bay checkerspot butterfly, *Euphydryas editha bayensis* : evidence for a metapopulation model. Amer. Nat. 132:360-382.

HELENTJARIS, T. 1987. A genetic linkage map for maize based on RFLPs. Trends in Genet. 3:217-221.

I.B.P.G.R. 1985. Cost-effective Long-term Seed Stores. International Board for Plant Genetic Resources, Rome.

KAHLER, A.L. ; M.T. CLEGG & R.W. ALLARD 1975. Evolutionary changes in the mating system of an experimental population of barley (*Hordeum vulgare* L.). Proc. Nat. Acad. Sci. USA 72:943-946.

KIMURA, M. 1983. The Neutral Theory of Molecular Evolution. Cambridge University Press, Cambridge.

LACAZE, P. 1986. Utilisation de marqueurs biochimiques dans l'appréciation de la variabilité génétique de variétés composites du blé. Mémoire de D.E.A. de Génétique des Populations. Université de Paris VII / Institut National Agronomique Paris-Grignon, Paris.

LANDE, R. & G.F. BARROWCLOUGH 1987. Effective population size, genetic variation, and their use in population management. In: SOULÉ, M.E. (ed.) Viable Populations for Conservation. Cambridge University Press, Cambridge.

LEVINS, R. 1970. Extinction. Pp. 77-107 in: M. GERSTENHABER (ed.) Some mathematical questions in biology. American Mathematical Society, Providence, R. I.

LOVELESS, M.D. & J.L. HAMRICK 1984. Ecological determinants of genetic structure in plant populations. Ann. Rev. Ecol. Syst. 15:65-95.

MARUYAMA, T. & P.A. FUERST 1985a. Population bottlenecks and nonequilibrium models in population genetics. II. Number of alleles in a small population that was formed by a recent bottleneck. Genetics 111:675-689.

MARUYAMA, T. & P.A. FUERST 1985b. Population bottlenecks and nonequilibrium models in population genetics. III. Genic homozygosity in populations which experience periodic bottlenecks. Genetics 111:691-703.

MCINTOSH, R.A. 1988. Catalogue of gene symbols for wheat. Pp. 1225-1323 in : T.E. MILLER & R.M.D. KOEBNER (eds.) Proceedings of the Seventh International Wheat Genetics Symposium held at Cambridge, UK, 13-19 July 1988. Vol. 2. Institute of Plant Science Research. Cambridge.

MERRIAM, G. 1988. Landscape dynamics in farmland. Trends in Ecol. and Evol. 3:16-20.

NEI, M. 1973. Analysis of gene diversity in subdivided populations. Proc. Nat. Acad. Sci. USA 70:3321-3323.

NEI, M. 1987. Molecular Evolutionary Genetics. Columbia University Press, New-York.

PAYNE, P.I. ; K.G. CORFIELD & J.A. BLACKMAN 1979. Identification of high-molecular-weight subunit of glutenin whose presence correlates with bread-making quality in wheats of related pedigree. Theor. Appl. Genet. 55:153-159.

PERNES, J. 1984. Gestion des Ressources Génétiques des Plantes. Tome 2 : Manuel. Agence de Coopération Culturelle et Technique, Paris.

PICARD, E.; C. PARISOT, P. BLANCHARD, P. BRABANT, M. CAUSSE, G. DOUSSINAULT, M. TROTTET & M. ROUSSET 1988. Comparison of the doubled haploid method with other breeding procedures in wheat (*Triticum aestivum*) when applied to populations. Pp. 1155-1159 in: T.E. MILLER & R.M.D. KOEBNER (eds.) Proceedings of the Seventh International Wheat Genetics Symposium held at Cambridge, UK, 13-19 July 1988. Vol. 2. Institute of Plant Science Research. Cambridge.

RICK, C.M. & J.I. YODER 1988. Classical and molecular genetics of tomato : highlights and perspectives. Ann. Rev. Genet. 22:281-300.

ROBERTS, E.H. 1975. Problems of long-term storage of seed and pollen for genetic resources conservation. Pp. 269-295 in: O.H. FRANKEL & J.G. HAWKES (eds.) Crop Genetic Resources for Today and Tomorrow. Cambridge University Press, Cambridge.

ROUSSET, M. ; J. KERVELLA, M. PICHON, G. BRANLARD, G. DOUSSINAULT & E. PICARD 1988. Effectiveness of recurrent selection for breadmaking quality breeding in wheat (*Triticum aestivum* L. em. Thell.). Pp.1009-1014 in: T.E. MILLER & R.M.D. KOEBNER (eds.) Proceedings of the Seventh International Wheat Genetics Symposium held at Cambridge, UK, 13-19 July 1988. Vol. 2. Institute of Plant Science Research. Cambridge.

TROTTET, M. 1988. Use of genic male sterility for breeding wheat lines resistant to *Leptosphaeria nodorum* Müller : results of a first selection cycle and prospects. Pp. 1199-1202 in: T.E. MILLER & R.M.D. KOEBNER (eds.) Proceedings of the Seventh International Wheat Genetics Symposium held at Cambridge, UK, 13-19 July 1988. Vol. 2. Institute of Plant Science Research. Cambridge.

VAN VALEN, L. 1973. A new evolutionary law. Evolutionary Theory 1:1-30.

VARVIO, S.L. ; R. CHAKRABORTY & M. NEI 1986. Genetic variation in subdivided populations and conservation genetics. Heredity 57:189-198.

WORLAND, A.J.; M.D. GALE & C.N. LAW 1987. Wheat genetics. Pp. 129-171 in: F.G.H. LUPTON (ed.) Wheat breeding. Its scientific basis. Chapman & Hall, London.

WRIGHT, S. 1931. Evolution in Mendelian populations. Genetics 16:97-159.

WRIGHT, S. 1969. Evolution and the Genetics of Populations. Vol. 2: The Theory of Gene Frequencies. The University of Chicago Press, Chicago.

WRIGHT, S. 1982. The shifting balance theory and macroevolution. Ann. Rev. Genet. 16:1-19.

Species Conservation: A Population-Biological Approach
A. SEITZ & V. LOESCHCKE (eds.) © 1991 Birkhäuser Verlag, Basel

Population Dynamics of Cory's Shearwater (*Calonectris diomedea*) and Eleonora's Falcon (*Falco eleonorae*) in the Eastern Mediterranean

D. Ristow[1]*, **F. Feldmann**[2], **W. Scharlau**[3], **C. Wink**[4] and **M. Wink**[4]*

[1] Pappelstraße 35, D-8014 Neubieberg, Germany;
[2] Universität Hamburg, Institut für Angewandte Botanik, D-2000 Hamburg, Germany
[3] Universität Münster, Mathematisches Institut, D-4400 Münster, Germany
[4] Universität Heidelberg, Institut für Pharmazeutische Biologie, D-6900 Heidelberg, Germany

Abstract

Colonies of Eleonora's falcons (*Falco eleonorae*) and of Cory's shearwaters (*Calonectris d.diomedea*) have been studied on an archipelago in the southern Aegean Sea since 1975. Ringing results show that young shearwaters have a remarkable degree of philopatry; especially males tend to breed in their natal or adjacent nest. Once settled, both sexes show a strong lifelong site tenacity and mate fidelity. Also Eleonora's falcons return almost exclusively to their natal colony in their second year. The retrap data of both species suggest that dispersal to and immigration from other colonies is rather low and as a consequence that the birds of each colony are genetically interrelated. These results are discussed in terms of gene flow and speciation. In addition other population parameters are described for both species (productivity, first estimates for juvenile and adult mortality).

Introduction

The genetic isolation of populations on islands has been considered a major event in speciation (MAYR 1976 , MACARTHUR & WILSON 1976, LACK 1947,1966). The numerous Aegean islands in the eastern Mediterranean often harbour colonies of two species, namely Cory's Shearwater (*Calonectris d.diomedea*) and Eleonora's Falcon (*Falco eleonorae*), which are almost restricted in their breeding distribution to the Mediterranean and the adjacent Atlantic islands. It would be interesting to know how these colonies are related and whether a tendency for adaptive radiation or subspecies formation is already visible.

Cory's shearwaters breed in the Mediterranean (*C.d. diomedea*), in the Cape Verde Islands (*C.d. edwardsii*) and in the Atlantic islands, such as Selvages, Madeira, the Azores

---

* To whom the correspondence should be addressed

and the Canary Islands (*C.d. borealis*). The shearwaters spend the winter months in the South Atlantic (CRAMP & SIMMONS 1977).

Cory's shearwaters return to their breeding grounds, usually small rocky islands, in March. Like other Procellariidae this shearwater breeds colonially. Nest burrows are often found under rocks (in the Mediterranean) or bushes (in the Selvage Islands). At the end of May the female lays a single egg which is incubated for 53 days. The young hatch in July and fledge in October (for review see (CRAMP & SIMMONS 1977, WINK *et al.* 1982, MOUGIN *et al.* 1984, ZINO *et al.* 1987). As compared to other Procellariidae knowledge on the breeding biology, population structure and demography of Cory's shearwaters is largely incomplete. The subspecies best studied is *C.d. borealis* from the Selvage Islands, because there the shearwaters breed in the open and can be caught quite easily, moreover the birds are rather abundant on these islands (JOUANIN *et al.* 1977, MOUGIN *et al.* 1984,1985, 1986, ZINO 1971, ZINO *et al.* 1987). The Mediterranean breeding colonies of Cory's shearwaters are comparably small, and usually situated in uninhabited rocky islands, which are difficult to visit. Furthermore, since many shearwaters are often inaccessible in their nest burrows , it is difficult to monitor a large number of birds and in consequence to study their population biology. For this reason, the number of studies on this subspecies is small and our general knowledge very scanty (FERNANDEZ 1985; MASSA & LO VALVO 1986; ZAMMIT & BORG 1987, 1988; RISTOW & WINK 1980; RISTOW *et al.* 1981; WINK & RISTOW 1979, WINK *et al.* 1979, 1982, 1987a,b).

The Eleonora's falcon is a unique raptor for several reasons. The species breeds colonially on rocky cliffs of the Mediterranean, on islands along the North African coast and in the Canary Islands (WALTER 1979). About 80% of the world population which consists of about 3000 pairs breed in Greece, especially in the Aegean islands. Eleonora's falcons are adapted to the autumn migration of Palaearctic birds to Africa in that they breed in the late summer and early autumn, a time which provides plenty of food to feed the young in terms of migrants.*F. eleonorae* returns from its wintering quarters in Madagascar in April and occupies its breeding grounds. Egg laying starts much later at the end of July, the young hatch after 30 d at the end of August, fledge after ca 35 d and leave their natal islands in October. Most falcon islands are difficult to reach, so that relatively few studies (WALTER 1979; RISTOW, WINK and coworkers) have been performed with this rare falcon.

We have studied the population biology of both species for many years. Research started in 1965 on a small archipelago in the southern Aegean sea [size 1 square km or less for each islet]. Since 1975 our research has been intensified with regular research activities every year. Several aspects of the physiology and population biology of both species have already been published (e.g. RISTOW 1975; RISTOW & WINK 1980, 1985, 1990; RISTOW *et al.* 1979,1980, 1981,1982, 1983a,b, 1986, 1989; WINK & RISTOW 1979, WINK *et al.* 1978, 1979a,b, 1980a,b, 1982a,b,c, 1985, 1987a,b).

In this study we evaluate the field data obtained between Aug. and Oct. 1988 and in June and Sept./Oct. 1985 and 1989, and summarize our earlier data in terms of population structure and population dynamics. We are aware that our ringing and retrap numbers are still relatively small and limited as compared to those of other species (gulls, terns, Manx Shearwater [BROOKE 1990]) which are usually much easier to trap/retrap because they normally live in more accessible, large colonies. Due to these technical difficulties our population data are certainly incomplete and some conclusions preliminary, but better data do not exist for *C.d.diomedea* and *F.eleonorae*.

Material and Methods

C. d.diomedea: It is difficult to ascertain the size of the shearwater colony studied (island size: 300 x 1000 m; its name and location are not given for reasons of conservation), but we estimate from the number of burrows found that about 1000 pairs are present. Typical distance between individual nests is 10-20 m but at locations of high density it can decrease to 3 m. A limited number of breeding sites and nest burrows are accessible, but it is often difficult to catch the birds in their burrows which lie hidden deep under rocks. We therefore concentrated our efforts on 2 study plots of 1000 m^2 (A) and 6000 m^2 (B) where burrows are only up to 2 m deep. In these plots we monitored all nest burrows, and especially in A we regularly ringed all adult birds and their respective young. In plot B, intensive ringing and retrapping were carried out in 1978, 1985 and 1989, meaning that about 100 nest sites were marked individually and monitored during egg-laying and incubation period. We tried to capture both partners of a breeding pair and ringed their nestlings later in the autumn. Between 1977 and 1988 723 fledglings and 470 adults were ringed. Field work was carried out by 2-4 persons/session (often in dark moonless nights) and overall we spent more than 250 days on the island studying Cory's shearwaters. Since we were living directly within the colony (plot A), we were able to record many behavioural and other aspects of the life of this nocturnal shearwater (e.g.RISTOW & WINK 1980; RISTOW *et al.* 1981; WINK & RISTOW 1979, WINK *et al.* 1979, 1982, 1987a,b). Our presence did not seem to disturb the birds, which returned to the same place sometimes every night, although we checked their ring numbers regularly.

F. eleonorae: There are about 600 pairs of Eleonora's falcons on the archipelago, with the largest colony of about 275 pairs on "falcon island". Between 1975 and 1989 about 2850 falcons were ringed, mostly as juveniles. Since 1980 each age-class has been marked with an individual, year-specific colour-ring. In 1983, 1985, 1986 and 1988 a substantial proportion of the breeding bird population was monitored with spotting scopes and the ring-combination of each bird was recorded. In 1984 and 1985 a similar control was carried out on an adjacent island, which is only 5 km away and can be seen from our "falcon island".

Results

Calonectris diomedea

Productivity

The productivity of *C.d.diomedea* was determined in 1985 and 1989 by monitoring 253 nests in June, the time of egg-laying and again in Sept./Oct., the time of fledging . Fledging young were scored as successful breeding, and failed eggs, dead chicks or empty nests (where an egg had been recorded before in June) as breeding failure: 17-19% of all eggs fail to hatch, and a further 4-5% of nestlings do not fledge. Thus breeding success is 77% on average (Tab. 1). Comparable figures have been recorded in other colonies: Malta 72% (ZAMMIT & BORG 1987), Marseille Islands 79-82% (FERNANDEZ 1985) and Selvage Islands 60-71% (ZINO 1971, ZINO *et al.* 1987). The lower productivity in the Selvage Islands may be due to an increased predation, since nest sites lie in the open there and are thus vulnerable to gulls and other predators.

Juvenile mortality

In a small subcolony of 12-14 pairs (within plot A) we ringed more than 90% of all nestlings between 1975 and 1989 (with the exception of 1978) and controlled the origin of the breeders in all of these years. Out of 62 fledglings, which were ringed between 1977 and 1984 (we have omitted fledglings from later years, since they were not yet due to return by 1989), we were able to recapture 14 different birds in later breeding seasons, especially in 1985 and 1989. All birds were sexed by their beak size dimorphism (WINK & RISTOW 1979; RISTOW & WINK 1980) and 93% were males (Tab. 2). These data can be used to estimate the order of magnitude of the survival probabilities of males between birth and the time of first return to their natal island: Assuming an equal sex distribution, we started with 31 males. 13 of them were recaptured at a mean age of 6 years (no attempt was made to correct for the differential age of the return of individuals because of the small sample size). A first and preliminary estimate for the geometric mean of the annual mortality probability of males is about 13% (m= 1 - $(13/31)^{0.167}$) or slightly lower as some males may have

Tab. 1:   Breeding success of *C. diomedea*. Marked nests were controlled in June and Sept./ Oct., both in 1985 and in 1989. Fledgings were scored as successful breeding, and failed eggs, dead chicks or empty nests as breeding failure, where an egg had been recorded in June before.

Year	Nests with success (%)	Failure (%)	Total number of nests
1985	121 (78.6%)	33 (21.4%)	154
1989	77 (76,2%)	24 (23.8%)	101

settled outside the control area and others born in 1983/84 may not have returned yet. The overall survival probability for young males between birth and first return is 41%. Female mortality is probably similar, but females show a much stronger dispersal (Tab. 2 and 3), which makes a separate calculation impossible at present.

Tab. 2: Recovery of Cory's shearwaters in their natal colony. In a subpopulation (plot A), which consisted of up to 14 pairs, all young were ringed each year between 1977 and 1989 (except 1978). They were later retrapped in the same study plot.

Year	Number of young ringed	Number of young controlled in later years		
		Males	Females	Total
1977	12	3	0	3
1978	0	-	-	-
1979	9	2	0	2
1980	9	2	1	3
1981	8	0	0	0
1982	9	0	0	0
1983	6	4	0	4
1984*	9	2	0	2
	62	13	1	14

*years 1985 till 1989 are omitted since no birds had been seen until 1989 which derived from this group.

Tab. 3: Distance between the natal nest and the site of first breeding or control in later years.

Distance (in m)	Breeding birds		"Bachelors"	
	Males	Females	Males	Females
0	3	0	1	0
0.5-5	2	0	5	0
5-20	3	0	3	0
20-50	1	0	1	0
50-100	2	3	0	1
>100	1	0	0	0
Total	12	3	10	1

Adult mortality

In 1978 and in 1985, 38 and 77 pairs of shearwaters were ringed respectively and again monitored on their former nests in 1985 and 1989 (Tab. 5). With respect to the high degree of site and mate fidelity (see below) we assume that birds which could not be recaptured in these years had disappeared and probably died in the meantime or were on "sabbatical" (MOUGIN et al. 1984, 1985). An approximation for the annual adult mortality (calculation as juvenile mortality) would be 16-17% for the 4- and 7- year assessments and 13% for females in the 1978 - 1989 comparison. A similar high value has been reported from Malta (ZAMMIT & BORG 1987), but is in contrast with data from other shearwaters, which all have adult mortalities below 10% (CRAMP & SIMMONS 1977). For *C.d. borealis* an annual mortality of 9.1% was obtained (MOUGIN *et al.* 1984). In the Selvage Islands it was shown that about 7% of all breeders interrupt reproduction and take an annual sabbatical leave (MOUGIN *et al.* 1985). We need to analyse whether Aegean shearwaters also take their "sabbatical" and whether it has influenced our mortality data.

In spring 1990 we have recaptured the birds at those nests which were occupied in 1989. Using these new data we arrive at an annual mortality of 10-11% and have no evidence for a significant "sabbatical" (RISTOW & WINK, in prep.)

If we analyse the fate of the 1978 birds which were still present in 1989 (see above), a remarkable bias towards females is evident. These data imply that females tend to become older than males (see also Tab. 4). Taking into account that *C. diomedea* first breeds at an age of 7 (Tab. 4), we can assume that males and females can live and reproduce up to an age of at least 20 years.

Tab. 4:   Age of first recovery, first breeding and maximal site tenacity and longevity of *C. diomedea*. **Age of first recovery**: Time between birth and recovery, i.e. a bird born in 1980 and recaptured in 1984 was considered to be 4 y old. **Age of first breeding**: Calculated similarly to first recovery. Birds had been ringed as fledgelings in the control area and were recaptured in the same area in a nest burrow in later years. (It is assumed that more females settled outside the control area). **Maximal site tenacity**: Evaluation of all recapture data from 1975 to 1989. Calculation: e.g. a bird breeding in the same nest in 1980 and 1981 had a site tenacity of 2 years. **Longevity**: Maximal time interval between capture and latest recapture of shearwaters which had been ringed as adult birds (with a typical age of at least 6 y). Calculation: A bird ringed in 1980 and recaptured in 1989 was assigned a longevity of 9 y; its real age was probably at least 9 + 7 = 16 y. **M** = males, **F** = females

	Number of Shearwater						
Years	First Recovery M+F	First Breeding M	First Breeding F	Max. Site Tenacity M	Max. Site Tenacity F	Longevity M	Longevity F
1	0	0	0	-	-	14	9
2	0	0	0	13	14	3	5
3	0	0	0	4	8	8	4
4	4	0	0	5	3	36*	35*
5	5	1	0	34*	24*	4	3
6	7	3	0	3	2	3	4
7	3	2	2	1	1	6	4
8	2	3	0	4	2	2	1
9	4	3	1	2	1	1	2
10	1	0	0	0	0	2	3
11	0	0	0	1	0	1	5
12	0	0	0	0	5	0	2
13	0	0	0	0	1	1	0
mean	6.4	7.3	7.6				

* biased by 1985/89 studies.

Tab. 5:   Estimation of annual adult mortality

Year of study	Number of breeders		Year of recapture	Number of recaptures		Annual mortality [%]	
	M	F		M	F	M	F
1978	38	37	1985	10	11	17	16
			1989	1	8	28	13
1985	72	77	1989	34	37	17	17

Age at First Breeding

We recaptured the first young birds at the age of four in their natal colony, similar to the situation in *C.d. borealis* (JOUANIN *et al.* 1977). The age of first recovery was 6.4 years on average (Tab. 4). First breeding was at 5 years for males and at 7 years for females, with an average of 7.3 y for males (n=12) and 7.6 y for females (n=3) (Tab. 4). In the Selvage Islands *C.d. borealis* breeds later for the first time with a mean of 9 years, but the authors do not distinguish sexes (MOUGIN *et al.*1984, 1986). On Malta a male (*C.d.diomedea*) was found breeding at the exceptional age of 2 years (ZAMMIT & BORG 1988).

Philopatry

Cory's shearwaters which were ringed in plot A and B have not been found breeding in other colonies (for instance on Malta or in Italy, where larger numbers of *C.d.diomedea* have been controlled). Only one of our shearwaters has been recovered away from the colony. It was a x+3-year old male and shot in Tunesia on migration in March. In contrast, of 280 fledgelings ringed in plots A and B between 1977 and 1984 26 were recaptured in their natal colony within the same area in later years, i.e. birds which were born in study plot A were later found in plot A but not in the adjacent plot B and vice versa. Fledglings from later years had not returned yet at the time of this evaluation (till 1989). Breeding sites up to 150 m away from the study plot A were monitored regularly and no recaptures of ringed birds (from A) were obtained in those areas.

We have analysed the dispersal patterns of the 26 recaptures (Tab. 3): 66% of breeding males show a remarkable degree of natal philopatry and were found either in their natal nest or within a 20 m radius of it, for instance, just in the neighbouring nest (Tab. 3). The same is true for male "bachelors" (These birds were caught while sitting at the entrance of a burrow of which the two breeding birds had been controlled inside the burrow previously. Compare ZAMMIT & BORG (1987) and MOUGIN *et al.* (1984)).This high degree of natal site tenacity has never been recorded before for Cory's shearwaters or - as far as we know - for any other Procellariidae (CRAMP & SIMMONS 1977). However, these data do not rule out that a small proportion of the males emigrate to other parts of the island which were not monitored.

The few females that were recaptured (Tab. 2 and 3) were observed 50-150 m away from their site of birth (4 birds, Tab. 3). These data suggest that most females will settle at greater distances, i.e. 150-500 m away from the study plots. Since we were not able to check these sites regularly, these females could not have been retrapped. For the bachelors controlled in later years when they were classified as breeders, again male birds were found much closer to their natal nest than female bachelors. Manx shearwaters (*Puffinus puffinus*) settle in their natal colony; in addition, some of the females (about 25%) but not the males, seem to disperse to adjacent colonies (HARRIS 1966, BROOKE 1978, 1990). It is not known yet whether the same applies to female Cory's shearwaters.

Site Tenacity

Once the birds have occupied their breeding site, they show a high degree of site tenacity: A number of birds returned to the same nest site for up to 11 years (males) or 13 years (females) (Tab. 4). Of 33 males and 37 females recorded breeding in 1985 and recovered in 1989 28 (=85%) and 27 (=73%), respectively, were still in the same nest burrow. The other birds (15% or 27%, respectively) were breeding within 50 m. Similar findings have been obtained for other shearwaters (CRAMP & SIMMONS 1977) and for *C. diomedea* (ZAMMIT & BORG 1987, MOUGIN *et al.* 1984).

Mate Fidelity

It could be a consequence of this remarkable site tenacity that Cory's shearwaters also show high mate fidelity: We could monitor a few pairs that had been breeding together for more than 8 years. In 1985 and 1989 we controlled mate fidelity in a quantitative way (Tab. 6): Out of 61 breeding pairs marked in 1985 only 20 (=32.8%) were present in 1989, of which both partners were still present. Of 23 pairs both partners (=37.7%) disappeared; of 8 and 9 pairs only the male or the female, respectively, was still present but mated to a new partner. About 95% of the 21 surviving pairs had maintained their partners for four years,

Tab. 6: Pair tenacity of *C. diomedea*
The fate of 61 pairs, which were breeding in 1985, was monitored in 1989.

	Number of pairs/ or birds	%
Breeding pairs in 1985	61	100%
Both partners disappeared	23	37.7%
Pairs together in 1989	20	32.8%
Pairs present but with different partners	1	1.6%
Only male present*	8	13.1%
Only female present*	9	14.8%
Total	61	

* with new mates

only one pair (=5%) had separated and was now breeding with their former neighbours. Lifelong monogamous pair bond seems to be the rule in Procellariiformes (CRAMP & SIMMONS 1977) and has been also recorded for *C.diomedea* (WINK et al. 1982, ZAMMIT & BORG 1987, MOUGIN et al. 1984).

Tab. 7: Age structure of an island population of *Falco eleonorae*. Young falcons were colour-ringed in each year between 1980 and 1989. In 1985, 1986 and 1988 the breeding population was intensely monitored for the presence of colour-ringed birds.

A. Comparison between the number of actually observed falcons (=obs) and the theoretical number of falcon which could be expected (=exp) if our population model was correct (calculations see RISTOW et al. 1989). Assumptions: Productivity is 0.6, mortality until first breeding 0.78, and adult mortality 13% (RISTOW et al. 1989). Population size: 275 pairs. From these data we have calculated, how many ringed falcons can be expected in each age class. M = males, F = females

Age   Numbers of falcons

| | 1985 | | | | 1986 | | | | 1988 | | | |
	M	F	total obs.	Exp.	M	F	total obs	Exp.	M	F	total obs.	Exp.
1	0	1	1		0	2	2	0	0	0		
2	6	11	17	18	2	1	3	4	9	10	19	28
3	9	6	15	18	2	2	4	5	10	14	24	25
4	6	8	14	16	0	3	3	5	6	3	9	9
5	14	7	21	21	1	2	3	4	7	6	13	11
6					3	4	7	6	8	1	9	11
7									5	2	7	10
8									6	8	14	13
not colour-ringed												
	101	104	205		34	27	61		78	69	147	
total number of monitored falcons												
	136	137	273		42	41	83		129	113	242	

B. Age class model, derived from Tab. 7 A. N = number of birds/ age class/ population of 550 falcons

Age (years)	1985		1986		1988		Average	
	n	%	n	%	n	%	n	%
2	64	12	59	11	49	9	57	10.4
3	52	9	50	9	61	11	55	10.0
4	46	8	34	6	61	11	47	8.5
5	46	8	33	6	56	10	45	8.2
6			51	9	35	6	43	7.7
7					26	5	26	4.7
8					35	6	35	6.3
Rest							242	44.2
								100.0

Falco eleonorae

Young Eleonora's falcons return to their natal island at the age of 2 to 3 years (exceptionally in their first year) and usually start breeding in their second year (females) or 2/3 years (males) (RISTOW *et al.* 1983; 1989). During their first year of life young birds generally do not return to their breeding grounds in the summer: Out of 16 birds whose rings were recovered outside the natal colony, 75% were found in their first year far away in the Mediterranean, such as Cyprus, Corsica or Malta. The other 4 recoveries were older and from the wintering grounds in Madagascar or as in the case of a 2-year-old non-breeding falcon, from Malta (RISTOW & WINK 1990). In contrast, several hundred controls or recoveries were achieved on the natal island (Tab. 7), indicating a high degree of philopatry. This assumption was supported by another approach: In 1984 and 1985 we controlled whether birds that had been born on "falcon island" were now breeding in a falcon colony (size about 200 pairs) on an adjacent island only 5 km away. We were able to check 59 (1984) and 149 (1985) breeding falcons, none of which was carrying a colour ring. Only a 2-year- old, colour-ringed female was spotted, which was probably not breeding. If a substantial immigration (e.g. 10%) had taken place by birds, which were born on "falcon island" (since this is the largest colony in the region, it would be likely that most immigrants would come from there) then the number of colour-ringed birds should have been in the order of 10 -30) and could not have escaped our attention. Therefore, these data imply that the falcons do not disperse to other colonies at a substantial degree (WINK *et al.* 1987). Once settled the falcons return to their nest territory every year and in some instances we could show that pair bonds were stable till the death of a breeding partner (RISTOW *et al.* 1979).

Earlier data showed a productivity of 0.6 young/adult, a juvenile mortality of 78% until first breeding and an annual adult mortality of 13% (RISTOW *et al.* 1985, 1989, WINK *et al.* 1987). Our additional data from 1988 (Tab. 7) are in agreement with the former productivity and mortality figures. Tab. 7 B implies that the annual adult mortality could be even somewhat lower, i.e. 11 or 12%.

Discussion

Although both species are taxonomically unrelated, some traits in their life history are similar: the birds return to their birth site and once settled, they show a very high degree of site tenacity and pair bond.

Since some of the young male Cory's shearwaters come back to the nest of their birth, we have to assume that even close relatives interbreed in the colony (for instance mother and son, if the former male partner has disappeared) which would increase the number of homozygotic alleles. In captivity, interbreeding of close relatives usually produces an enhanced number of deformed or otherwise damaged offspring. We speculate that the same happens on this island but that the affected offspring do not survive. On the other hand the homozygotic birds have the advantage of maintaining gene-encoded specializations or adaptations to their particular breeding site and population structure, which would be lost easily if birds from outside with different genes or alleles were to enter the colony regularly. At present we are analysing the degree of relatedness within and between shearwater populations with the aid of DNA fingerprinting (WINK *et al.* 1990).

Allowing enough time and no change in the dispersal behaviour we speculate that this and other shearwater populations will develop into subspecies or species eventually, as described for the Darwin finches on the Galapagos archipelago (LACK 1947). For *C. diomedea* several subspecies are already distinguished, which differ in size (s. above). Within the nominate group, *C.d. diomedea*, we observe a size gradient from west to east, with the shearwaters in the Aegean being the smallest birds (M.Wink, D.Ristow, H.Witt, in prep.; MASSA & Lo VALVO 1986). This implies either that the island populations of the Mediterranean are separated for a prolonged period or a strong selection pressure must exist which maintains the size gradient, or more likely both factors are responsible. It has to be discussed whether the shearwaters from different parts of the Mediterranean already deserve the status of specific subspecies at the present time because of their distinctive size differences.

The entire population of Eleonora's falcons winters in Madagascar and in a small part of East Africa, where the birds from different colonies are probably in near contact. We could expect this to lead to the interchange of members of different populations. However, our ringing data do not reveal such a trend. On the contrary, they clearly demonstrate a high degree of philopatry and site tenacity.

As the Eleonora's falcon is a relatively rare species and specimens in museums are limited in number, nobody has attempted to distinguish its populations or subspecies, although differences are likely to exist [e.g. the proportion of falcons of the black and the light colour phase differs between colonies (WINK *et al.* 1978); clutch size is 3-4 in the western and 2-3 in the eastern populations (WINK *et al.* 1985)]. Again, DNA analysis, which is in progress in our laboratory (WINK *et al.* 1990) will probably help to solve this question in the future.

A high level of philopatry is not only typical for island birds, but has also been reported for many mainland passerine species; e.g., young birds of *Ficedula hypoleuca*, *Parus major*, *Delichon urbica*, *Hirundo rustica* etc. tend to settle near the sites of their birth when first breeding (BERND & STERNBERG 1968, LACK 1966, NYHOLM 1986). To some degree the mainland populations behave like island populations and the importance for speciation is likely to be similar (MAYR 1967, OBDAM *et al.* 1984).

Both species have about the same adult mortality, but *F. eleonorae* invests its reproductive effort in a higher number of offspring with a corresponding higher juvenile mortality. On the other hand in Cory's shearwaters productivity is much lower but the survival probability until first breeding is substantially higher. According to results presented for other avian species, this might mean that for Eleonora's falcons a smaller number of females contribute a much larger fraction of the next generation, suggesting that relatedness between individuals could be different for the two species.

Although we have invested much effort in the study of both species during the last two decades, our data are still small as compared to those of other species whose colonies are easier to reach and thus easier to monitor. Nevertheless, we think that our data, although sometimes preliminary, will contribute to the understanding of the "secret life" of Eleonora's falcons and Cory's shearwaters. But there are still many problems and mysteries to be solved in the biology and population genetics of both species.

Zusammenfassung

Brutkolonien des Eleonorenfalken (*Falco eleonorae*) und des Gelbschnabelsturm-tauchers (*Calonectris d.diomedea*) wurden auf einer Inselgruppe der südlichen Ägäis seit 1975 intensiv untersucht. Die Ringkontrollen zeigten, daß junge Sturmtaucher eine auffallende Geburtsortstreue aufweisen: besonders die Männchen siedeln sich in ihrem Geburtsnest oder einem nah gelegenen Nachbarnest an. Sobald die Sturmtaucher einmal zu brüten begonnen haben, zeigen beide Geschlechter eine bemerkenswert hohe Brutorts- und Partnertreue. Auch die Eleonorenfalken brüten später fast ausschließlich auf ihrer Geburtsinsel. Diese Daten lassen vermuten, daß bei beiden Arten ein Dispersal und ein Genfluß von oder zu anderen Inseln gering ist. Demnach müssen die Angehörigen einer Inselpopulation relativ eng miteinander verwandt sein. Mögliche Konsequenzen für die Bildung von Unterarten und Arten werden diskutiert. Zusätzlich erfolgen Angaben zur Produktivität, der Jugend- und Adultmortalität und zur Altersstruktur beider Arten.

Acknowledgements

We thank our Greek and German friends for help during field work, especially Jutta Ristow, Till Ristow, Astrid Scharlau, and Helmut Ludewig and an anonymous referee for his helpful comments.

References

BERNDT,R. & STERNBERG, H. (1968): Terms, studies and experiments on the problem of bird dispersion. Ibis 110:256-294.

BROOKE, M. de L. (1978): The dispersal of female Manx Shearwaters *Puffinus puffinus*. Ibis 120: 545-551.

BROOKE, M. (1990): The Manx Shearwater. Poyser, London.

CRAMP, S.,& SIMMONS, K.E.L. (1977): Handbook of the birds of Europe, the Middle East and North Africa. Oxford Univ. Press, Oxford, London, New York.

FERNANDEZ,O. (1985): Etude synoptique des observations relatives au nid du Puffin Cendre *Calonectris diomedea* sur les Isles de Marseille. Alauda 53, 147-148.

HARRIS,M.P. (1966): Age of return to colony, age of breeding and adult survival of Manx Shearwater, *Puffinus puffinus*. Bird study 13:84-95.

JOUANIN, C., ROUX, F. & ZINO, A. (1977): Sur les premiers resultats du baguage des Puffin Cendres *Calonectris diomedea* aux Iles Selvagens. L'Oiseaux et R.F.O. 47: 351-358.

LACK, D. (1947), Darwin's Finches. Cambridge University Press, Cambridge.

LACK, D. (1966): Population studies in birds. Clarendon Press, Oxford.

MACARTHUR, R. & WILSON, E.O. (1976), The theory of island biogeography. Princeton Univ. Press, Princeton.

MASSA, B & M. lo VALVO (1986) Biometrical and biological considerations on the Cory's Shearwater, *Calonectris diomedea*. NATO Adv.Ser., Vol 612, 293-313.

MAYR, E. (1976), Evolution and diversity of life. Harvard University press, Cambridge

MOUGIN, J.-L., JOUANIN, C. & ROUX, F. (1985): Donnees complementaires sur les années sabbatiques du Puffin Cendre *Calonectris diomedea borealis* de l'Ile Selvagem Grande. Bocagiana 86,1-12.

MOUGIN, J.-L., JOUANIN, C., DESPIN, B. & ROUX, F. (1986): The age of first breeding of Cory's Shearwater on Selvagem Grande and problems of ring loss. Ringing & Migration 7:130-134.

MOUGIN, J.-L.., ROUX, F., JOUANIN, C. & STAHL, J.-C. (1984): Quelques aspects de la biologie de reproduction du Puffin cendre *Calonectris diomedea borealis* des Iles Selvagens. L'Oiseau et R.F.O. 54:229-246.

NYHOLM, N.E.I. (1986): Birth area fidelity and age of first breeding in a northern population of Pied Flycatcher *Ficedula hypoleuca*. Orn. Scand. 17:249-252.

OBDAM, P., VAN DORP, D., & TER BRAAK, C.F.J. (1984): The effect of isolation on the number of woodland birds in small woods in the Netherlands. J. Biogeography 11:473-478.

RISTOW, D. & WINK, M. (1980): Sexual dimorphism of Cory's Shearwater. Il-Merill 21:9-12.

RISTOW, D. & WINK, M. (1990): Distribution of non-breeding Eleonora's Falcon. (in press).

RISTOW, D. (1975): Neue Ringfunde vom Eleonorenfalken. Vogelwarte 28:150-153.

RISTOW, D., & WINK, M. (1985): Breeding success of the Eleonora's Falcon and conservation management. In: (Newton,I., Chancellor,R.D., eds) Conservation studies on raptors. ICBP Techn. Publ. 5:147-152.

RISTOW,D., CONRAD,B., WINK, C.,& WINK, M. (1980): Pestcide residues of failed eggs of Eleonora's Falcon from an Aegean colony. Ibis 122:74-76.

RISTOW,D., SCHARLAU, W. WINK, M. (1989): Population structure and mortality of Elonora's Falcon, *Falco eleonorae*. In: Raptors in the modern world (eds. Meyburg, B.U., Chancellor, R.D.) WWGBP, Berlin, London,Paris, pp 321-326.

RISTOW,D., WINK, C. & WINK, M. (1979): Site tenacity and pair bond of Eleonora's Falcon. Il-Merill 20:16-18.

RISTOW,D., WINK, C. & WINK, M. (1981): Telemetrie der Körpertemperatur des Gelbschnabelsturmtauchers (*Calonectris diomedea*). Vogelwelt 102:57-60.

RISTOW,D., WINK, C. & WINK, M. (1982): Biology of Eleonora's Falcon. 1. Individual and social defense behaviour. Raptor Research 16:65-70.

RISTOW,D., WINK, C. & WINK, M. (1983): Biologie des Eleonorenfalken (*Falco eleonorae*). 12. Die Anpassung des Jadgverhaltens an die vom Wind abhängigen Zugvogelhäufigkeiten. Vogelwarte 32:7-13.

RISTOW,D., WINK, C. & WINK, M. (1986): Assessment of Mediterranean autumn migration by prey analysis of Eleonora's Falcon. In: (A.Farina, ed.) First Conf. on birds wintering in the Mediterranean region, 285-295.

RISTOW,D., WINK, C. ,WINK, M. & FRIEMANN, H (1983): Biologie des Eleonorenfalken (*Falco eleonorae*). 14. Das Brutreifealter der Weibchen. J.Orn. 124:291-293.

WALTER, H. (1979). Eleonora's Falcon. Adaptation to prey and habitat in a social raptor. Chicago Univ. Press. Chicago.

WINK, M. & RISTOW, D. (1979): Zur Biometrie des Sexualdimorphismus des Gelbschnabelsturmtauchers (*Calonectris diomedea*). Vogelwarte 30:135-138.

WINK, M., RISTOW, D. & WINK, C. (1979): Biologie des Eleonorenfalken (*Falco eleonorae*).3. Parasitenbefall während der Brutzeit und Jugendentwicklung. J.Orn. 120: 64-68.

WINK, M., RISTOW, D. & WINK, C. (1985) Biology of Eleonora's Falcon. 7. Variability of clutch size, egg dimensions and egg colouring. Raptor Res. 19:8-14.

WINK, M., RISTOW,D., & SCHARLAU, W. (1987): Niedrige Ei- und Körpertemperatur (Hypothermie) bei brütenden Gelbschnabel-Sturmtauchern (*Calonectris diomedea*). J. Orn. 128:334-338.

WINK, M., SWATSCHEK, I., FELDMANN, F., SCHARLAU, W., & RISTOW, D. (1990): Untersuchungen von Verwandtschaftsbeziehungen in Vogelpopulationen mittels DNA-Fingerprint. Die Vogelwelt 111: 86-95.

WINK, M., WINK, C. & RISTOW, D. (1979): Parasitenbefall juveniler und adulter Gelbschnabelsturmtaucher (*Calonectris diomedea*). Bonn. Zool. Beitr. 30:217-219.

WINK, M., WINK, C. & RISTOW, D. (1982): Brutbiologie mediterraner Gelbschnabelsturmtaucher *Calonectris d. diomedea*. Seevögel, Sonderband, 127-135.

WINK, M., WINK, C. & RISTOW,D. (1978): Biologie des Eleonorenfalken (*Falco eleonorae*).2. Zur Vererbung der Gefiederphasen (hell-dunkel). J.Orn. 119:421-428.

WINK, M., WINK, C. & RISTOW,D. (1980): Biologie des Eleonorenfalken (*Falco eleonorae*). 8. Die Gelegegröße in Relation zum Nahrungsangebot, Jagderfolg und Gewicht der Altfalken. J. Orn. 121:387-390.

WINK, M., WINK, C. & RISTOW,D. (1980): Biologie des Eleonorenfalken (*Falco eleonorae*). 9. Eitemperaturen und Körpertemperaturen juveniler und adulter Falken. Vogelwarte 30: 320-325.

WINK, M., WINK, C. & RISTOW,D. (1982): Biologie des Eleonorenfalken (*Falco eleonorae*). 10. Einfluß der Horstlage auf den Bruterfolg. J.Orn. 123: 401-408.

WINK, M., WINK, C. & RISTOW,D. (1982): Biologie des Eleonorenfalken (*Falco eleonorae*).11. Biometrie des Sexualdimorphismus adulter und flügger Falken. Vogelwelt 103:225-229.

WINK, M., WINK, C., SCHARLAU, W. & RISTOW, D. (1987): Ortstreue und Genfluß bei Inselvogelarten: Eleonorenfalke (*Falco eleonorae*) und Gelbschnabelsturmtaucher (*Calonectris diomedea*). J.Orn. 128:485-488.

ZAMMIT, R.C. & BORG,J. (1987): Notes on the breeding biology of the Cory's Shearwater in the Maltese Islands. Il-Meril 24, 1-9.

ZAMMIT, R.C. & BORG,J. (1988): Cory's Shearwater breeding in its 2nd year. Il-Meril 25, 11.

ZINO, P.A. (1971): The breeding biology of Cory's Shearwater *Calonectris diomedea* on the Selvage Islands. Ibis 113:212-217

ZINO, P.A., ZINO, F., MAUL, T., & BISCOITO, J.M. (!987): The laying, incubation and fledgling periods of Cory's Shearwater *Calonectris diomedea borealis* on Selvagem Grande in 1984. Ibis 129: 393-398.

Species Conservation: A Population-Biological Approach
A. SEITZ & V. LOESCHCKE (eds.) © 1991 Birkhäuser Verlag, Basel

Housing Viable Populations in Protected Habitats: The Value of a Coarse-grained Geographic Analysis of Density Patterns and Available Habitat

C. Schonewald-Cox [1] and M. Buechner [2]

[1] National Park Service, Cooperative Studies Unit, Institute of Ecology, Division of Environmental Studies, Wickson Hall, University of California, Davis, CA, USA
[2] Division of Environmental Studies, University of California, Davis, CA, USA

Abstract

While a multi-disciplined approach is necessary to restore or protect threatened populations, it is difficult to achieve. This is especially true for mammalian carnivores. Considerable attention is payed to small population requirements for maintaining natural variation and obtaining genetic and demographic viability. There are several subject areas we stress as being crucial to conservation of species such as for carnivores. We stress three of these in this manuscript: 1) Use of space (species' large and small-scale area requirements for viable populations) 2) Habitat capacity for protection (Consequences of habitat fragmentation, including within reserves) and 3) Extinction trajectories and density patterns that influence and are effected by population dynamics. We conclude by stating some changes in the breath of planning that would assist in future conservation efforts of species such as for our case example, the mammalian carnivores.

Introduction

Current events such as anthropogenic encroachment and disturbance result in increased habitat fragmentation, and reduce the capacity of landscapes to provide for the viability of wildlife. In many areas wildlife populations may be reduced to such an extent that they fall below the tolerance limits, both demographically and genetically. Such species require immediate attention in order to (1) determine the survival requirements for a viable, "self-sustaining" population (2) ensure that adequate protection policy exists for in situ conservation of the species (e.g., ensure the availability of protected habitat and the enforcement of regulations); and (3) utilize ex situ techniques if necessary to provide for the temporary continuation of the species until points 1 and 2 can be achieved.

While protection in situ is ideal, it gives us the least control over the propagation, genetic changes and demographic fluctuations of the population, and allows relatively little direct access to information on these processes. For species or populations facing imminent threat of local extinction, such control and information may be vital (e.g., Black-footed ferrets, red wolves, grizzly bears or timber wolves in the continuous 48 U.S. states, etc.). Captive studies have assisted in wildlife conservation by providing an assessment of genetic and demographic requirements for some species and can provide for the maintenance of the last remnant populations of some species (RALLS, *et al.*, 1986). In addition, field techniques for monitoring individual movements and reproduction, and packaged demographic models have recently become easier to access. As a result, we are closer to defining what constitutes viability or self-sustainability in a population.

However, a knowledge of the genetic and demographic viability requirements of a population is not sufficient in most circumstances for direct field application. In order to ensure the viability of a species in situ, especially in the absence of active management, we must be able to provide sufficient protected habitat area (SCHONEWALD-COX, *et al.*, submitted). This, in turn, requires an understanding of the spatial patterns (e.g., dispersal and dispersion) of wildlife populations. The widely varying requirements of species result in a need to understand space utilization patterns at variable geographic scales (e.g., CARLILE, *et al.*, 1989; GOSZ & SHARPE, 1989; TURNER, *et al.*, 1989; WEINS & MILNE, 1989). For large predator populations, the space (i.e., area) required is likely to be on the order of tens of thousands of square kilometers.

Is space available in our human-dominated habitats to maintain viable or self-sustaining populations of carnivores? In many cases we are left to rely on any remaining, and currently protected, natural habitat and have no alternative measures to assist declining populations (e.g., restoration of habitat). In addition, habitat fragmentation is continuing in most areas (SCHONEWALD-COX & BUECHNER, 1991). These trends will often eliminate the opportunity for long-term field studies of space utilization for 'relatively' undisturbed wildlife. We need to develop means to assess the space requirements of the population at variable scales of geographic resolution, quickly. Techniques for the analysis of how choice os spatial scale affects interpretation of animal space utilization are available and have been utilized across small areas (e.g, WEINS & MILNE, 1989) and in larger landscapes (TURNER, *et al.*, 1987) .

How much space is required is not purely based on expected density or the amount of "useable habitat" but is complicated by the effect of intrusive impacts that erode habitat quality; these are influenced specifically by the development and type of boundary conditions occurring on the reserve and the fragmentation and landuse patterns of adjoining land (SCHONEWALD-COX & BAYLESS, 1986; SCHONEWALD-COX, 1988; SCHONEWALD-COX & BUECHNER, 1991). Anthropogenic effects such as these tend to reduce available resources even within otherwise useable space.

Studies of the space requirements of animals have been directed mostly to small habitat scales, such as are typically represented in reports of density (no/km^2) or home range size (km²/indiv). Even when larger censuses have been conducted, they tend to focus on areas of less than 10^3 km² (Figure 1). A habitat size of 10^3 km² is a relatively small area in terms of

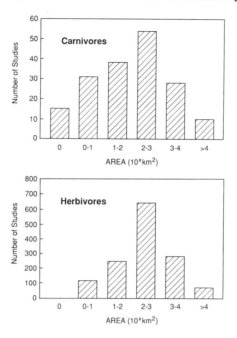

Fig. 1: The number of census studies conducted over the last 50 years in areas of given sizes. Based on a review of census information for (a) large carnivores and (b) large herbivores (mainly ungulates); details reported in SCHONEWALD-COX, *et al.*, (in press).

landscape dynamics at a regional scale or at a scale ideally required for conservation of large bodied carnivores (10^5-10^6 km²; such as for wolves, big cats). In order to apply what we have learned from the study of behavior, genetics and demography, to estimate the minimal space requirements for a viable population or the maximum capacity of a reserve for population size, we also need to test and model large-scale spatial parameters.

Estimating large-scale space requirements:

Carnivores as an example

The Carnivora lie at the top of the trophic pyramid as predators and scavengers. Some are also omnivorous. They often represent umbrella species whose conservation requirements are likely to encompass those of many other components of their communities simply due to their large space requirements (e.g., area requirements such as for foraging requirements, etc., GITTLEMAN, 1989; EWER, 1973; EISENBERG, 1981). Large carnivores are among the first to show obvious effects of habitat loss and environmental disturbance. However, because of their low densities, crepuscular or nocturnal feeding habits, olfactory orientation, and cryptic nature, most carnivores are difficult to study. The accumulation of baseline population data is difficult and expensive. Conservation of carnivore species is a slow process6. If we were to wait for sufficient knowledge on carnivore habitat requirements to be available before developing some guidelines on management of carnivores, we might watch these species pass by all the more quickly. Just as we need broad guidelines on demographic and genetic requirements for species´ viability, we also need them on ecological and geographic requirements.

In spite of the problems associated with carnivore conservation, there has been considerable behavioral, physiological and ecological study conducted on the more familiar species at small geographic scales. Course- scale geographic and ecological information is available for a few species. The availability of the products of this broad-scale ecological work makes it possible for us to conduct a preliminary examination of censuses in which we look for trends in the possible area requirements and habitat use of carnivore populations (at multiple spatial or geographic scales). By examining the combination of these together with information on genetic and demographic aspects of carnivore population viability and effects of habitat fragmentation, we may be able to develop successful conservation strategies.

We (SCHONEWALD-COX et al., in press) recently conducted a study in which we began to address the subject of space utilization by medium and large-sized mammalian carnivores in the Families Felidae, Canidae, Mustellidae, Hyenidae, and Ursidae. We examined how carnivores were reported to utilize space at geographic scales larger than at home range size or reduced to density per km². Our data were collected from studies conducted in differing habitats and under a variety of habitat conditions, including those affected by anthropogenic disturbance.

We focused upon the relation between the reported number of individuals and the area of the census site keeping a focus on habitat area rather than density per single average km². We also examined the variation in reported density relative to both census site area and to female body weight for purposes of comparison with other published works. Our sample sizes were seriously affected by the preponderance of home range-related and small-scale density reports which we could not use. Census number-study site area regressions were significant for *Panthera leo*, *Panthera tigris*, *Crocuta crocuta*, *Meles meles*, *Mephitis mephitis*, *Ursus americanus*, *Canis lateans* and two different types of *Canis lupus* populations (the wolf regressions are shown in Figure 2). Other species, for which samples were too small to regress census number against area, appeared in the significant trophic and social type regressions. Our general carnivore regression which included all studies in the sample was also weakly significant. While sample sizes were less than ideal, our carnivore survey provided sufficient data to suggest that there may be real trends in how carnivores are spread out in habitat areas ranging from a few hundred to $10^{4}+$ square kilometers. But it became evident to us that effects of scale and census technique were probably affecting many of our regressions.

As might be expected, densities decline with increasing area (SCHONEWALD-COX, *et al.*, in press). In our carnivore regressions we also found a disproportionately large variation among habitat sizes in the ten's of km² compared to larger areas; but samples were also smaller for large areas. Censuses over large habitat areas tend to include more transient individuals than small areas, and due to the geographic limits of the census, more ranges of quality than do small scale studies, potentially affecting density estimates. The detection of population trends is problematical. For example, reported wolf densities have declined over

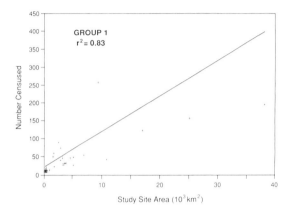

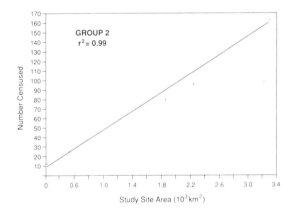

Fig. 2: Regression of wolf numbers versus area based on a review of wolf census data from studies conducted over the last 50 years (for details of the review see SCHONEWALD-COX, *et al.*, in press). Wolf populations reviewed could be separated into two groups on the basis of the number/area regression, (a) a low density group and (b) a high density group (note different scales on axes of graphs). Both regression lines are significant and account for large proportions of the variance in wolf numbers; the slopes of the two regression lines are significantly different.

the last 50 years (Figure 3), but it is not clear whether this is the result of study methods, a shift toward studying less dense populations as time progresses, or an actual decline in wolf densities over time.

Our analysis suggests that densities derived from conventional small-scale studies of space use (e.g., behavioral studies measuring home range size) are insufficient to adequately describe carnivore space requirements at the large geographical scales for many species (SCHONEWALD-COX, et al., submitted). In addition to the home range size or density of individuals within small areas, data on factors such as the degree of overlap between home ranges, patterns of home range tenure, the number and spacing of individuals which do not occupy stable home ranges, the amount of area not utilized regularly by resident individuals, and the patterns of seasonal or yearly deviations from standard within-home range movements are needed if we are to specify large-scale patterns from small-scale data. In addition, variability in home range based densities is likely to be disproportionately higher than that of census-based densities, because the subset of home range samples is small within any study population, and measurement criteria remains subjective in the literature. Because the estimated capacity of a habitat is pivotal to the provision of viability requirements for populations, viability assessments should include a comparison of space use patterns measured at various scales.

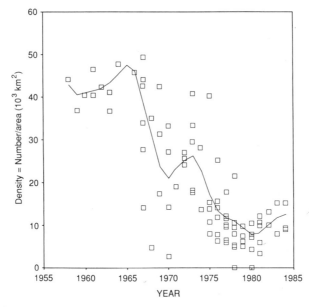

Fig. 3:   When all annual observations of density from the census studies of wolves reviewed in SCHONEWALD-COX, et al. (submitted) are plotted versus the year in which the study was reported, a substantial downward trend is obvious. The line shown is a broken-stick nonparametric regression fitted by computer. The cause of this trend is unknown; it may result from changing census methods, from a trend toward studying less dense populations in recent years, or from an actual decline in wolf population densities over time.

SALWASSER, *et al.* (1987) published some preliminary carnivore data and compared park sizes to the space required by carnivores and found tentatively that they would probably be too small to accommodate large carnivores. Most of the available areas of protected habitat which could potentially accommodate carnivore species (e.g., National Parks) are less than 10^4 km² in area (Figure 4). Unfortunately, even areas nearly an order of magnitude larger than this (e.g., Yellowstone National Park) appear to be too small to support self- sustaining populations of the largest carnivores into the future (e.g., SHAFFER, 1981). A comparison of park sizes shows that most parks are under 10^2 - 10^3 km², areas too small to support large carnivore populations. The effects of severe landscape fragmentation, such as that commonly found in Europe and increasingly in North America, will act to exacerbate this trend and to increase the total area necessary to sustain large predators. Cooperation between parks and adjacent lands, such as national forests, seems essential for accommodating viable populations of large carnivores. But the policies specifying the nature of a productive cooperation with adjacent federal lands (as in the U.S.) still need to be developed (SCHONEWALD-COX, *et al.*, submitted).

The consequences of habitat fragmentation

Once we know at what geographic scale we must plan for protection of self-sustaining populations for given species or cohort we must recognize another major problem that affects space requirements, that of major landscape fragmentation in protected habitats. Landscape fragmentation generally refers to the subdivision of formerly continuous community or ecosystem by humans. The description of landscape fragmentation processes originated in Europe where its consequences are best understood. In Europe, several thousand years of colonization and road building have resulted in increasingly dense subdivision of the landscape. The probability of providing undivided, unchallenged, and undisturbed space to species that require 10^4 or more km² of land to maintain a minimal viable population is essentially non-existent except in the most inhospitable climates and terrains.

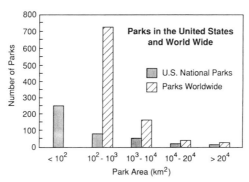

Fig. 4:   Areas covered by National Parks or equivalent reserves both in the United States and worldwide. Areas taken from FRANKEL & SOULÉ (1980) and from U.S. National Park Service data.

Habitat fragmentation is spreading in North America. Where it occurs, fragmentation of landscapes reduces space available to some large and vulnerable species including carnivores. The term "fragmentation" implies a partial to total isolation of populations. Landscape fragmentation may not cause population fragmentation for species capable of moving across barriers (e.g., roads) in the landscape matrix. However, while landscape fragmentation is not equivalent to population fragmentation, its disruptive effects may be substantial for some species. For example, an increase in fragmentation which does not effectively isolate subpopulations results in increased amounts of "edge" within the landscape, both reducing the habitat available for patch interior species and allowing the entry of edge-associated disturbances.

Currently, we rely on parks to provide protection from fragmentation effects. However, as far as parks are concerned, landscape fragmentation is not exclusively an external phenomenon. In one study, presently underway (SCHONEWALD-COX & Buechner, 1991), we have initiated an analysis of internal fragmentation within existing parks by paved roads. This is pertinent to the determination of space requirements for carnivores, e.g. for determining the population capacity of the park. If park fragments function together as a single unit, the space requirements and capacity have one set of values (in terms of N_e/km^2. However, if the park functions as several, semi-isolated units, space requirements may be underestimated and park capacity overestimated. We have initiated the study by asking how much park land is lost to roads and the potential edges generated by them, and what geometric (including spatial) effects result from this fragmentation. There is considerable literature on impacts entering parks. In our pilot study we found that the large parks of the lower 48 states (not including Alaska or Hawaii), as well as small ones, are substantially fragmented by paved roads. The roads tend to cut through the center of the parks and to divide them into several fragments. Most of the parks which were analyzed in the pilot study have between 2 and 10 fragments. The upper limit to the number of fragments within parks falls steadily with park size with the exception of Death Valley and Yellowstone National Parks. Small parks had more fragments than large parks. The majority of fragments are small with most falling below 10^2 km^2. Few of the fragments are greater than 10^3 km^2. We estimated that most of the parks surveyed would lose little protection area from road effects extending less than 250 meters on either side of the roads. However, effects extending 500 meters or more into a park will affect substantial fractions of the total area for parks surveyed. Furthermore, except in the largest of these parks, the majority of park areas lie within 10 km of paved roads. Medium to large carnivores, have daily movements of >>10 km^2. Disturbance associated with road traffic and impact could increase the total area required by single individuals living near roads and reduce the carrying capacity of the park.

Habitat fragmentation may have beneficial effects for some populations. Fragmentation can increase the structural diversity of the landscape matrix and increase the variability between fragments. Provided there is high connectivity between major patches, the predator may find the new habitat provides more feeding and denning opportunities than previously;

it may provide more room for other social and age classes than before. However, the increase in disruption of animals moving within their traditional home range spaces could decrease the efficiency with which they extract resources, and increase their exposure to impacts, ultimately decreasing survival.

Fragmentation may obstruct small scale catastrophic problems. Depending on the degree of autocorrelation between habitat changes in different areas of the landscape, this obstruction may increase the ability of the habitat to support viable populations of focal species. However, the barriers to catastrophe are potentially barriers to gene flow. They can constrict home range size, and slow demographic recovery, increasing the N_e/MVP required to maintain the species in the habitat. For less mobile species, barriers predispose the establishment of metapopulation units that each have to meet individual requirements for a local level of stability relative to the migration rate. In the worst cases, geneflow and recovery of population subunits can be manipulated by management.

In Europe, and increasingly in North America, we are faced with planning for conservation in highly fragmented landscapes in which remnant areas of natural habitat are relatively small. In these cases, we may be able to protect large vertebrates such as carnivores in clusters of relatively small fragments if we can (1) protect the animals from intrusive effects moving into the patches from the adjoining areas (e.g., SCHONEWALD-COX, 1988) and (2) artificially compensate for decreased gene flow between population fragments in isolated fragments of habitat. In such cases an estimate of the space required at large geographic scales is important because the heterogeneity that exists over the range of remnant patches is likely to be high due to (1) the social, including spacing, behavior of the focal species; (2) natural (biogeographic) variation in habitat variables such as soil type, local climate, or vegetation patterns, and (3) human modifications of the landscape in and around remnant habitat patches. It will be important to investigate the influence of scale on our recognition of population and habitat features over the landscape of interest. It is possible, even probable, that the area required to support a species in a heterogenous set of patches may be much larger than that estimated from straight-line extrapolations of data from small geographic scales when their habitats are impacted by intrusive generated edge effects. In addressing the subject of population viability, serious attention must be paid to the relationship between space requirements and habitat fragmentation.

Extinction trajectories and density patterns

As the fragmentation and alteration of natural landscapes continues, sensitive species face both the loss and the deterioration of required habitat, processes result in the increasing rarity of sensitive species. (RABINOWITZ, et al. 1986) defined seven forms of rarity affecting the flora of the British Isles; their organization includes geographic range, habitat specificity and local population size. Here, we will consider the population trajectories followed by species which are becoming increasingly rare due to the loss or alteration of available habitat.

Species which are decreasing due to habitat changes may become rarer in two senses: (1) restriction of geographical range, and (2) reduction in density (for the purposes of this paper, we will assume that the habitat specificity of the species does not change). Direct loss of habitat results in the constriction of the geographical range of a species. Alternatively, the alteration of habitat by anthropogenic effects can reduce the density of a population in some or all parts of its range. Changes in density and/or geographic range may lead to the decline and, in some cases, eventual extinction, of local populations (RABINOWITZ, et al., 1987; GILPIN, 1987). Various combinations of density and range changes are possible as a species or population declines. Four such "extinction trajectories" are outlined below and illustrated in Figure 5.

(A) **Density reduction.** The population may continue to be found across the same geographical range while its density decreases throughout that range. This may result from habitat deterioration which reduces the capacity of the habitat to support the species at high densities but which does not cause the actual destruction of the habitat. Regional effects such as acid rain, climate changes, and the release of toxins into the environment may result in this type of decline in a species. Alternatively, the direct removal of animals from the

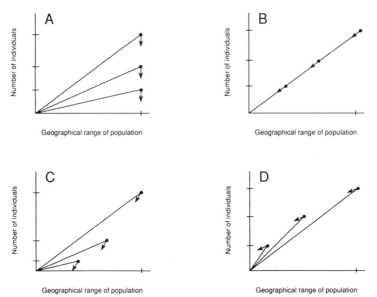

Fig. 5:   Some possible density and range patterns followed by species which are declining due to habitat loss and alteration or other disturbance (e.g., hunting). A: Decline in geographical range while density remains unchanged. B: Reduction in density while overall geographical range remains unchanged. C: Simultaneous reduction in range and density, and D: Declining range while density increases in the remnant habitat of the species. In all cases the overall number of animals (Ntotal) is assumed to be decreasing.

populations, e.g., by fishing, hunting or poaching, may produce a range-wide decline in density.

(B) Range restriction. The geographical range of the population may decrease while the original density is maintained. This is probably the most common trajectory for species which face ongoing habitat loss but which persist in relatively pristine habitat within protected areas.

(C) Combined range and density reduction. For species facing both the outright loss of habitat and the deterioration of any remnant habitat areas, both the geographical range and the density of population may decrease. This "double danger" is expected to produce the most rapid decline of populations. Populations which begin their decline with simple density reduction or simple range restriction may later fall into this category as impacts multiply or extinction vortices take over (e.g., GILPIN & SOULÉ, 1986).

(D) Concentration. Increasing anthropogenic impacts may effectively concentrate the remaining individuals of a local population into small remnant areas. If the habitat quality remains high in the remnant areas, the species may decline across its former range while increasing in density in the few protected areas remaining relatively undisturbed. This scenario is most likely when the boundaries of protected areas act as reflecting barriers for park animals or for species in which normal population regulatory mechanisms, e.g., predation or emigration, are absent or blocked in the remnant habitat areas.

Such patterns may not be evident from studies which focus on small areas; studies of the large-scale space-utilization patterns are vital if we are to elucidate patterns of decline. Our work with carnivores is an example of how one might establish a baseline against which such changes in density and/or range can be analyzed, when short-time limits our chances of success.

Conclusions

Space limitations, anthropogenic disturbances, the internal disruption of protected areas, and other factors likely to induce extinction are nearly certain to occur more rapidly than many of our conservation efforts can act to counter them, simply because of the time required for the planning and actions necessary for long term conservation. Planning for the protection of threatened species requires information on behavior and population ecology gathered at several geographic scales, the careful description of spatial and temporal population trends and a synthesis of such information with policy and management objectives as outlined in Figure 6, rather than focusing narrowly on genetic and demographic viability. Ideally, declining species would be studied in detail (e.g., censuses, detailed population information) while sufficient numbers remain that conservation efforts can be carefully planned years in advance. Unfortunately, for many species we have neither the requisite time nor funds for such efforts; in some critical cases, we simply do not have enough animals or habitat left for such studies. In these cases, rather than develop new field data, we may focus on reviews of currently available census data analyzed with careful attention to the effects of geographic scale. For those species for which available census data show consistent trends, we may be able to engage in some advance planning necessary for conservation objectives. Even for species for which data is scarce or less reliable, we can use reviews of information at broad geographical scales to establish estimates of baseline space requirements, population trends, and the probable effects of habitat fragmentation.

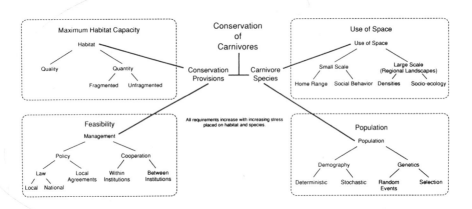

Fig. 6:   An overview of the levels of organization needed to address the conservation of carnivores.

Zusammenfassung

Zur Wiederherstellung bzw. zum Schutz gefährdeter Populationen ist ein multidisziplinärer Ansatz erforderlich. Eine erfolgreiche Umsetzung ist aber schwierig, insbesondere für carnivore Säuger. Besonders die Erfordernisse kleiner Populationen im Hinblick auf ihre genetische und demographische Lebensfähigkeit sowie das herkömmliche Management freilebender Tierarten werden herausgestellt. Zusätzlich können zahlreiche andere Gesichtspunkte beim Schutz großer Raubtiere entscheidend sein. Drei davon werden in diesem Manuskript vertieft behandelt:

1)   die groß- und kleinräumigen Anforderungen der Arten an ihren Lebensraum zur Aufrechterhaltung lebensfähiger Populationen;

2)   Konsequenzen von Habitatzerteilungen - auch innerhalb von Schutzgebieten;

3)   Aussterbewahrscheinlichkeiten und Dichtemuster.

Zum Schluß werden einige Änderungen des Planungsverfahrens dargelegt, die geeignet sind, um zukünftige Schutzmaßnahmen für große Raubtiere zu begleiten.

References

CARLILE, D.W., J.R. SKALSKI, J.E. BATKER, J.M. THOMAS & F.I. CULLINAN. (1989): Determination of ecological scale. Landscape Ecology 2:203-213.

EISENBERG, J.F. (1981): The Mammalian Radiations. University of Chicago Press, Chicago.

EWER, R.F. (1973): The Carnivores. Cornell University Press, Ithaca, NY.

FRANKEL, O.H. & M.E. SOULÉ. (1981): Conservation and Evolution. John Wiley, New York.

GILPIN, M.E. (1987): Spatial structure and population viability. Pp. 125-140 in M.E. Soulé (ed.) Viable Populations for Conservation. Cambridge University Press, Cambridge, 189 pp.

GILPIN, M.E. & M.E. SOULÉ. (1989): Minimum viable populations: processes of species extinction. Pp. 19-34 in M.E. Soulé (ed.) Conservation Biology: The Science of Scarcity and Diversity. Sinauer Associates, Sunderland, MA, 584 pp.

GITTLEMAN, J.L. (ed.) (1989): Carnivore Behavior, Ecology and Evolution. Cornell University Press, Ithaca, NY.

GOSZ, J.R. & P.J.H. SHARPE. (1989): Broad-scale concepts for interactions of climate, topography, and biota at biome transitions. Landscape Ecology 3:229-243.

RABINOWITZ, D., S. CAIRNS & T. DILLON. (1986): Seven forms of rarity and their frequency in the flora of the British Isles. Pp. 182-204 in M.E. Soulé (ed.) Conservation Biology: The Science of Scarcity and Diversity. Sinauer Associates, Sunderland, MA, 584 pp.

RALLS, K. & J.D. BALLOU (eds.). (1986): Proceedings of a workshop on genetic management of captive populations. Zoo Biology, volume 5.

SALWASSER, H., C. SCHONEWALD-COX & R. BAKER. (1987): The role of interagency cooperation in managing for viable populations. Pp. 159-173 in M.E. Soulé (ed.) Viable Populations for Conservation. Cambridge University Press, Cambridge, 189 pp.

SHAFFER, M.L. (1981): Minimum population sizes for species conservation. Bioscience 31:131-134.

SCHONEWALD-COX, C. (1988): Boundaries in the protection of nature reserves. Bioscience 38:480-486.

SCHONEWALD-COX, C. & J. BAYLESS. (1986): The boundary model: A geographic analysis of design and conservation of nature reserves. Biol. Cons. 38:305-322.

SCHONEWALD-COX, C. & M. BUECHNER. (1991): Parc protection and public.roads. In FEIDLER, P. L. & S. JAIN (eds.) Conservation Biology: The Theory and Practice of Nature Conservation, Preservation and Managemant. Chapman and Hall, New York.

SCHONEWALD-COX, C., R. AZARI & S. BLUME. in press. Space-use at geographic scales and the conservation of carnivores.

TURNER, M.G. (1987): Landscape Heterogeneity and Disturbance. Springer Verlag, NY.

TURNER, M.G., V.H. DALE & R.H. GARDNER. (1989): Predicting across scales: Theory development and testing. Landscape Ecology 3:245-252.

WEINS, J.A. & B.T. MILNE. (1989): Scaling of 'landscapes' in landscape ecology, or, landscape ecology from a beetle's perspective. Landscape Ecology 3:87-96.

Are There Minimal Areas for Animal Populations ?

M. Mühlenberg, T. Hovestadt and J. Röser,

Ökologische Station der Uni Würzburg, OT Fabrikschleichach, D - 8602 Rauhenebrach, Germany

Abstract

The minimal area of an animal populution is determined by A) the area requirement of reproductive units and B) by the viable population size. A) varies due to individually different and seasonally fluctuating home range sizes and is in addtiion strongly influenced by habitat quality. Population survival depends on deterministic as well as stochastic events and can therefore be estimated only with limited probability. A certain limitation of risk factors can be achieved by enlargement of the population size, increase in number of suitable habitats and reduction of isolation between inhabited areas. To determine the size of a "minimum viable population" (MVP) a "population vulnerability analysis" (PVA) is used as most important data base. The objectives of a MVP (e.g. 95% survival probability for the next 100 years) determines the necessary environmental conditions. A method which allows faster predictions was developed for special demands in practical implementation.

A target species should be selected to give qualitative reasons for the protection of areas. Criteria for the selection of target species for conservation were developed which should be modified according to regional conditions.

The concept of target species can also be used to quantify the evaluation of habitats scientifically, which is also an important step for management practises.

The analysis of the data for selected species demonstrates the high variability of the area requirements, above all due to different habitat quality. For a MVP the area requirement is much higher than so far assumed (e.g. much more than 10 km^2 for a passerine species). It is not possible to create a generally valid catalogue for the area requirements of species.

1. Introduction: conservation strategies

When analyzing current nature conservation politics, two strategies which do not exclude but support each other can be distinguished: a) strategy of limitation of impact, and b) creative species conservation.

The strategy of **limitation of impact** aims primarily at the reduction of environmental impacts, usually on large areas. Determination of acceptable levels of environmental pollution and the promotion of less intensive agriculture are the most obvious goals of this strategy. Besides these large scale measurements to stop the man-made alterations of natural areas, local actions can be taken as well. These can be actions to prevent the immediate extinction of endangered species or to protect SPECIES DIVERSITY (conservation of "genetic resources"), in most cases by creating some form of nature reserve. The protected areas, however, cannot definitively prevent further extinction of species caused by fragmentation, environmental alterations, and general deterioration of environmental conditions (WILCOX 1984). They may slow down the loss of biological diversity; these areas may also function as a risk insurance against adverse forms of human land use and provide reference areas for research on populations.

The goal of **creative species conservation** is to improve the persistence of species. The main aim is to enlarge the chances for the survival of species. Basically, the society decides what time frame (protection for 50, 100 or 1000 years) and what level of protection will be selected (see chapter 4). Measures cannot be taken for all species and they would not be generally valid, either. Therefore TARGET SPECIES have to be chosen (see chapter 5). Special measures for protection of single species do not make sense if the conditions for the survival of communities are not met (see ELLENBERG *et al.* 1989). The concept clearly aims at the protection of communities but the success of actions will be considered with respect to the target species selected.

While practical actions for the protection of target species may be similar to those stated in the paragraph above, there is an important difference: all measures are taken (for scientific reasons) to benefit the selected target species. The aim of this article is to demonstrate that usually only a strategy for the protection of target species allows for participation of scientists in the formulation of protective measures, that it helps to clarify conflicting goals in nature conservation and that we need clear political statements about the principal goals of conservation politics.

2. The current status of nature reserves

At present, nature reserves in Germany have not been selected for and do not correspond to the demands of long-term sustainable populations. Consequently, they often cannot fulfil the requirements for effective protection (HAARMANN 1979):

-   they are too small for animal populations with their particular dynamics in space and time,

-   they are too isolated to enable exchange with other populations or re-colonization after local extinction,

-     they do not provide the specific requisites of the populations for which they have been
      created.

    Although the conditions of most nature reserves is in no way satisfactory (see
PLACHTER 1985), at present, we cannot do without them since our methods of land use
threaten the maintenance of the diversity of animals and plants. However, nature
conservation and thereby species conservation cannot be restricted to areas within the
narrow confines of nature reserves: the whole area of Germany is habitat for wild animals
and therefore should be called upon in all planning policies. The creation of a nature reserve
cannot mean that all other areas should be used without consideration of conservation
requirements (BURKEY 1989; see also ERZ 1981, PLACHTER 1984), such a policy would be
doomed to fail.

    With respect to the limitation of deterioriation of all areas by environmental pollution
(e.g. soil protection by reduction of pesticides and eutrophication), we have to ask how
much reduction in agricultural intensity is necessary to achieve species protection and
continuity. Assuming that further agricultural surplus production in Germany is avoided in
accordance with current EC policy, we can calculate the actual reduction in agricultural
production rather exactly (HAMPICKE 1988): a reduction of 10-20 % of current production.
Compared to the diversity of moderately cultivated land, HAMPICKE believes that this
amount is not enough to ensure protection of sensitive species.

    So, what size of area is necessary for the protection of species? This is the most
important question in nature conservation since it is well known that large areas can support
more species than small areas. There is no real substitute for the protection of natural
habitats unless we are satisfied with keeping animals in wildlife parks or in zoos. In the
following paragraphs we discuss some of methods conceived to answer this question.
Chapter 4 introduces a possible way for determining area requirements of self-sustaining
populations.

3. Can we know the minimal areas for ecosytems?

    One of the best documented ecological relationships is the (loglinear) correlation
between area size and number of species. This is the case for real islands as well as for
habitat islands, (e.g. number of large mammals on mountain tops of the rocky mountains
(PICTON 1979), arthropods (MADER & MÜHLENBERG 1981) or bird species (OPDAM &
SCHOTMAN 1987) in forest islands and in swamp areas (BROWN & DINSMORE 1986)).

    On a very small scale this species-area correlation does not seem to be valid at all or
only with different coefficients. The home range sizes of many species are not reached and
the characteristics of the areas are influenced mainly by external factors (MADER 1983). For
spiders and carabids, MADER (1981a) even found an inverse relationship between area and
total number of species. This seems to be a consequence of the edge effect, which means

that small habitat islands will increasingly be colonized by species not typical for the respective habitat (see USHER 1989, Fig. 1). ∅

Various hypotheses have been advanced to explain the correlation between area size and the number of species. The species-area relationship can be interpreted as a simple statistical phenomenon: since a larger number of individuals occur in large areas, the probability of rare species being represented increases (see "passive-sampling hypothesis" (CONNOR & MCCOY 1979) and "random-placement hypothesis" (COLEMAN *et al.* 1982)). According to the dynamic equilibrium hypothesis (GGH) of MACARTHUR & WILSON (1963), the number of species is a result of an equilibrium between extinction and immigration of species. The larger the distance between an island and a source of colonizers, the lower the immigration rate of new species; the smaller an island, the more likely species will become extinct due to the population size. Finally, in many cases, habitat diversity allows an equally good or better prediction of the number of species than the size of the area per se (BOECKLEN & SIMBERLOFF 1986). To a certain degree habitat quality could compensate for size, but usually a reduction of area size is associated with a decrease in habitat quality. There is evidence for all of these hypotheses and actually all hypotheses may be correct.

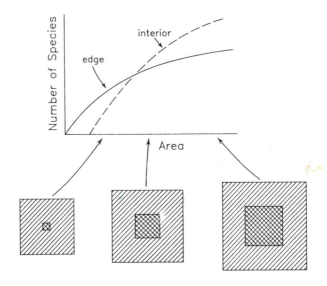

Fig. 1:   The effects of increasing the area of a census plot on the relative areas of edge and interior habitat (below) and on the rates of accumulation of "edge" and "interior" bird species in the census (above).

In the middle of the 70s it was debated whether more species can be maintained in a single large area than in two or more areas adding up to the same size (Single Large Or Several Small, **SLOSS**). The aim of this discussion was to find an optimal configuration of conservation areas (WILCOX & MURPHY 1985). BURKEY (1989) explains the intensity of the SLOSS discussion partly with the existence of different aims in nature conservation. Supporters of large reserves usually want to keep the rate of extinction as low as possible after the isolation of the area. Those who favour the alternate strategy (several small areas) want to cover as many species as possible. Nowadays the whole discussion has basically been put aside: usually the species-area relationship does **not** guide our attempts to find the optimal size and distribution of nature reserves.

However, as a result of this discussion some applications appeared, which can be of use in the formulation of conservation strategies. Based on the empirically derived species-area relationship, we can recognize areas especially diverse in species and select these sites for conservation priorities (USHER 1980; REICHHOLF 1980). Also, under the assumption that the remaining areas will not be influenced or changed, we can estimate how many species can be maintained in areas of a certain size (KANGAS 1987 SIMBERLOFF 1987) or how many species will be lost after a reduction in size. The loss of species takes places even in large areas. Since their isolation from the mainland, Borneo has lost 20 of its original 51 mammal species, Sumatra 8 of 51 and Java 26 of 50 species (FRANKEL & SOULÉ 1981).

As a rule of thumb, the number of species is doubled by a tenfold enlargement of an area (DIAMOND & MAY 1976). Accoding to this formula about 75 % of the remaining species in West Germany will become extinct since nature reserves make up only 1% of the country. Probably this is too pessimistic as many species live in other areas as well, but on the other hand it may be too optimistic since this 1 % is not one large area but is divided into many small patches which have only little value, especially for the endangered species (HAARMANN 1983 PLACHTER 1985). Based on the total area of nature reserves we will probably keep more than 25 % of all species but less than 25 % of the endangered species.

The characteristics of habitats change with a reduction in size, especially due to increasing external, mostly anthropogenic effects on the habitat island (JANZEN 1983; SOULÉ 1986). For plants in forest islands, an edge effect extending up to 15 m has been found (RANNEY et al. 1981), for arthropods even up to 40 m (MADER 1979). THIOLLAY (1989) studied several large game birds in South American rainforest reserves and found that their population density was still affected by hunting which took place up to 20 km outside the protected area. Locally, habitat fragmentation often does not lead to a decrease in number of species but to changes in species composition. Usually habitat generalists are favoured. Predation pressure becomes more severe in small forest islands and along forest edges (MARTIN 1988; WILCOVE 1985; BLAKE & KARR 1987; WHITCOMB et al. 1981; MAYFIELD 1977; ANDREN & ANGELSTAM 1988; TEMPLE & CARY 1988). Such an effect can still be demonstrated at a distance of 300-600 m from the forest edge (WILCOVE 1985). In small habitat fragments in the USA this is the main reason for loss of passerine species (WILCOVE et al. 1986) in addition to the nest parasitism by brown-headed cowbirds

(MAYFIELD 1977; BRITTINGHAM & TEMPLE 1983). Habitat fragmentation often leads to changes in species interactions. Also relations between plants and their pollinators are weakened and the plants produce fewer seeds (JENNERSTEN 1988).

For a long time, the continued existence of ecosystems or communities has been the aim of nature conservation. This approach was based on the assumption that every animal can only survive in a habitat suitable to its requirements. Furthermore, this aim is also based on characteristics of ecosystems. These are, among others, nutrient cycles, steady state, self regulation, stability or diversity of species. But is it possible to derive a minimum area requirement quantitatively from such characteristics?

None of the current definitions of an ecosystem allows a clear spatial delineation defined by **biological properties** of the system. Ecosystems are open systems. We do not want to make the "ecosytem" concept sound dubious although it is often no longer mentioned in new books on ecology (ANDERWARTHA & BIRCH 1989; BEGON *et al.* 1986). We do not think that the existing definitions are suitable as a basis for investigations of area requirements in respect to effective protection of ecosystems or species diversity. According to our knowledge, ecosystems - at least those in temperate zones - do not collapse in a catastrophic manner but change into a new state with different characteristics and species inventory. In any case, area size can only be examined with particular functions or benefits of ecosystems in mind. Such functions could be air filtration by a forest, water supply for a town or protection against avalanches in the mountains. Actually these aspects often play an important role in establishing reserves (e.g. water shed protection areas). In such areas, nature conservation is often not the main aim, rather direct benefits to man or even purely economic reasons predominate.

In sum, neither the species-area relationship nor current knowledge about ecosystems allow for a general statement about the surface area necessary for nature conservation. In any case it will be necessary to agree on which animal group should benefit from the protection measures.

4. Determination of area requirements of animal populations

The simplest (empiricism) and most often used way to determine area requirements of animal populations in practical conservation is based on the assumption that the area requirements of any one species are fulfilled where we actually find the species. However, island ecology as well as population biology and experience tell us that this assumption is completely unjustified. What other ways are there to define area requirements of sustainable populations?

Certainly, a direct approach to estimating the size of a long-term viable population would be to do appropriate **experiments**. It is necessary to create areas of different degrees of isolation and size and to monitor the development of the populations in these areas (SHAFFER 1981; LEMKUHL 1984; MÜHLENBERG & WERRES 1983). For statistical validity, multiple repeats are necessary. While this approach is straightforward, it has considerable drawbacks: it takes a very long time, it requires very large areas for experiments and it is very expensive. Especially the time factor prohibits a general solution of the area problem using an experimental approach.

Island ecology provides another approach to the problem. For a particular species, a species-area curve can be drawn which represents its **probability** of occurrence on an island of a certain size (DIAMOND 1978). Diamonds's definition of a "minimum area" is the size of an area on which a particular species can be found with an arbitrarily defined probability (e.g. 80 %). The probability for occurrence of a species on an island becomes higher with increasing size of the island (Figs. 2 and 3; see also figures in VAN DORP & OPDAM 1987). For several species, the probability curve rises steeply with increasing size. This seems to be especially true for very mobile and common species which colonize an area as soon as its size meets the requirements of a breeding pair or another social unit. In this case, a population consists of individuals of several islands or the whole archipelago.

Areas which are **inhabited** by a species may not necessarily suffice to **support** this species in the long term. Since most investigated islands are situated near continents, their species may occur only because of permanent re-colonization from the sources. Thus, the area size necessary for the long term persistence of species cannot be based on incidence curves alone. They can help, however, in determining the minimum size of areas which are **acceptable** to species and can therefore assist in conservation planning.

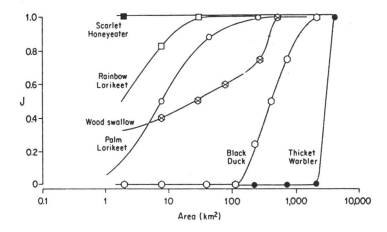

Fig. 2:   Incedence curves for 6 bird species from the Archipelago of New hebrids in the SW-Pacific. (DIAMOND 1978)

Currently there seems to be only one way to calculate the area necessary for the protection of a specific species: gather relevant information about the biology of the species of interest and then estimate the areal requirements of a population. We have compiled the following data for selected species:

1)   recent occurrence and abundance in West Germany and neighbouring countries;

2)   population development (within the last decades);

3)   area requirement: home range for the whole life cycle and its relation to habitat quality;

4)   demographic data;

5)   interactions with other populations.

Table 1 summarizes estimates of area requirements and the parameters influencing them in selected species. Such a table may be of help to planners in landscape design and nature conservation, who wish to obtain the required data for their expertises on regulations of human environmental impacts or habitat evaluation in the form of a catalogue (like in column "area requirement for about 100 pairs" in table 1). Due to the high variability of factors the data comprise a wide range. For planning purposes, only an arbitrary selection could be made which cannot be the goal of a scientific contribution. Therefore we emphatically object to the use of such data for planning purposes.

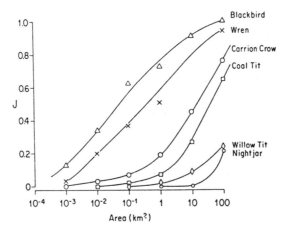

Fig. 3:   Incedence curves of 6 bird species from the British forest islands valid for the breeding season. (DIAMOND 1978)

Tab. 1: Data of home ranges, population densities, and dispersal distances for calculating the area
requirements of well studied species. See text for possible misuse of such informations.
(A= animal; P = pair; Fam = family group)

Species	Home range	Density	Dispersal distance	Area requirement for about 100 pairs	Notes
	Depending on age, sex, social status, seasonal fluctuations, habitat quality	Habitat quality, degree of isolation, predation	Dependant on time, age, population density; extreme or mean (probability) values	N_e (effective population size), social units, viable population size	On special features affecting population dynamics
Nuthatch	1 - 4 ha	1 - 6 A / 10 ha	? up to 100 km	≈ 1000 ha	habitat quality
Middle W. Spotted Woodpacker	10 ha	0.2 - 18 / 10 ha	< 10 km	≈ 2000 ha	very low dispersal
Magpie	5 - 10 ha			10000 -50000 ha	density dependent on Carrion Crow and Goshawk
Sparrow Hawk	10 - 3500 ha	40-80 P/100km²	1 - 265 km	≈ 20000 ha	temp. and rain April DDT
Little Owl	10 - 3500 ha	0.5-1 P/100 km²	(Ø up to 30 km)	≈ 20000-30000 ha	
Eagle Owl	1.2 - 2000 ha	1P/80-100 km²	? 40 km	≈ 1.000 000 ha	
Grey-lag Goose		1-2 P/100 ha		≈ 10000 ha	combination of diff. habitats susceptable to breeding disturbance
Chamois	100 - 300 ha		high dispersal	> 6000 ha(20 R)	demographic stochasticity
Red Fox	400 - 1600 ha	0.2-2 Fam/km²	3 - 100 km	≈ 20000 ha	rabies + control 60 % loss
Stoat	2 - 250 ha	3-10/100 ha	fast dispersal	≈ 5000 ha	prey dependent fluctuations of density
Wall lizard	5 - 60 m²	1 A/3-40 m²	< 0.1 km	2 - 5 ha	all populations isolated
Scarce Swallowtail	100 - 200 ha	2 A/ ha	2 - 3 km	50 - 600 ha	

However, even the rough estimates provided in Table 1 show that a population of
passerine birds, for example, requires several thousand hectares, a larger area then the great
majority of the nature reserves in Germany. From a biological point of view, it is not

possible to give fixed area sizes ensuring the survival of particular animal species. The remaining uncertainty in the data and the probability estimates in the various columns are mainly based on the following aspects:

1) **Habitat quality** and therefore resource availability vary in space and time. Animals depending on these resources will adapt their home ranges and population densities within wide limits.

2) data on **dispersal distance** require a definition of the time scale. There is some confusion on whether only distances between juvenile dispersal should be compared or whether movements of adults are to be included. Usually it is not known whether recorded animals contributed to the genetic exchange between subpopulations or whether they founded new subpopulations.

3) the main question is for which **number of individuals** the area requirements should be calculated; it cannot be answered in general terms or without detailed study. It can be answered scientifically only within certain probability limits based on risk analyses of different populations. For this, analysis information about population dynamics, especially the yearly fluctuations characteristic for all populations are of primary importance (see MÜHLENBERG 1990). The problems involved will be treated in detail in the following chapter.

4.1 The concept of "minimum viable populations" (MVP)

The object of conservation is to counteract the extinction of populations. In general, small populations will become extinct faster than large ones and populations in small areas will disappear sooner than those inhabiting large areas. This is clearly demonstrated by the consequences of progressive reduction and isolation of suitable habitats (habitat fragmentation) which is nowadays probably the most important reason for loss of species (TERBORGH 1974; SOULÉ 1983; SALWASSER et al. 1984).

The reason for the extinction of populations may be the alterations of habitat, nowadays usually by man, but also as a result of natural (chance) processes. Conservationists try to prevent change of habitat at least in some areas, but only a complete understanding of the causes of extinction will allow for well-founded conservation politics.

The concept of *"minimum viable populations"* (MVP) allows a quantitative investigation of area requirements:

A MVP for a certain species in any given habitat is the smallest isolated population with a defined chance (e.g. 95 %) of persistence over a specific period of time (e.g. 100 years) despite the foreseeable effects of demographic and genetic stochasticity on the populations as well as environmental fluctuations and natural catastrophes (after SHAFFER 1981).

The *minimum viable population (MVP)* size must be estimated in order to determine for how many social units the area requirements should be calculated. To prevent local extinctions, it is necessary to multiply the number of areas and consider the distance between the habitats (the likelihood of recolonization). All this is needed to estimate the total **area requirement of a viable population** (see fig. 4). Area requirement depends on home range of the individuals and habitat quality. We want to determine the necessary number of individuals for long term survival. The size of MVPs depend on risk factors like demographic and genetic stochasticity, environmental fluctuations, and catastrophes. An example for risk factors leading to the extinction of populations is shown in figure 5.

Within the scope of a "**population vulnerability analysis**" (PVA) we can investigate the importance of these risk factors in conjunction with the ecological situation and biological characteristics of the species concerned. We can use a risk analysis to estimate the survival probability of a population under current circumstances or certain modifications thereof. We can also find out the size of a population (and the necessary ecological conditions) fulfilling the MVP-criterion. To avoid tedious risk analysis, we can estimate the survival probability of a population to some extent from its observed characteristics. Important parameters are the population dynamics, the population structure (local distribution) as well as basic information on population growth and dispersal behaviour (see chapter 6).

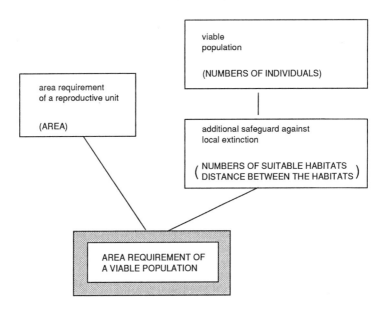

Fig. 4:   Information required about areal demande of a reproductive unit (individuum, pair, group) and size of viable population to calculate total area for protection of a population. Additional areas are included to protect more than one population.

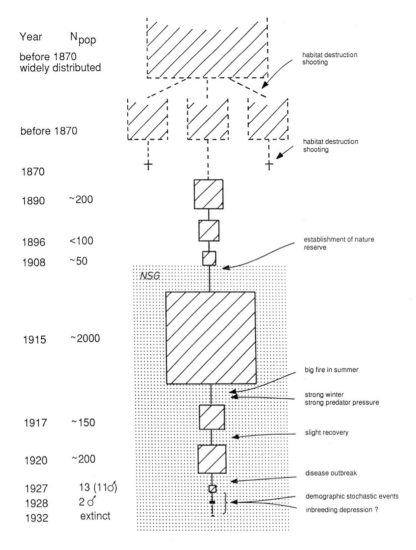

Fig.5:   Fate of the population of east-american heathhen (*Tympanuchus cupido cupido*) since
1870 on Martha`s Vineyard. In 1908 a guarded conservation area was established to
protect this last population. Severals events or processes (arrows) had such disasterous
effects on the population that extinction in 1932 was the final result. Originally the
species was common and well distributed in all of eastern North-America. (after
SIMBERLOFF 1986)

But how can the size of a MVP be determined? The process leading to a conclusion
can be outlined in 6 steps:

step 1

We can observe that natural populations are subject to considerable fluctuations in individual numbers (fig. 6). We know that populations may become extinct as a consequence of such natural fluctuations. The degree of fluctuation is usually lowest for large vertebrates and highest for insects (up to factor 100) . Small mammals (factor 10) or passerine birds (factor 3) have a medium position (MÜHLENBERG 1990). We want to define the mean population size and the conditions (size and quality of area, type of habitat, habitat fragmentation) which allow the persistence of a population with a defined probability (e.g. 95%) over a specified period of time (e.g. 100 years): a MVP.

step 2

We are interested in the specific reasons for decrease or increase of a particular population (arrows in fig. 6). In order to recognize these, we depend on field studies on the

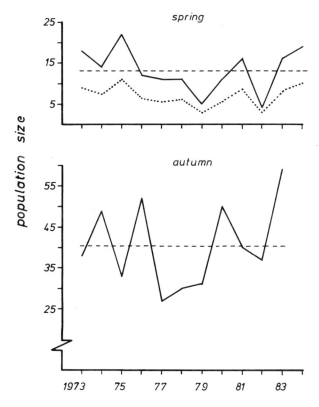

Fig. 6:   Census of a swedish population of nuthatches (*Sitta europaea*). Data on population size were taken in spring (top) and fall (bottom). Dotted line indicates number of females only (from NILSSON 1987).

ecology and demography of the population and a review of data from the literature. As an example, important determinants for the population dynamics of Swedish nuthatches (*Sitta europaea*) seem to be winter temperature and availability of beech-nuts in autumn. In the case of the Californian checkerspot butterfly (*Euphydryas editha*), the dynamics of subpopulations up to the extinction of local populations is influenced by the amount of summer precipitation, slope exposition, and food plant supply (WEISS *et al.* 1988). In Franconia, for an explosive increase in cricket populations some successive, extremely warm summers are sufficient (REMMERT 1984).

step 3

To estimate the survival probability under certain environmental conditions, e.g. for a nuthatch population, we would have to monitor about 500 different nuthatch populations

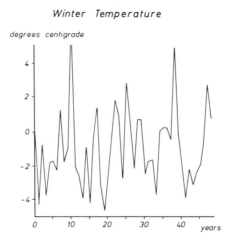

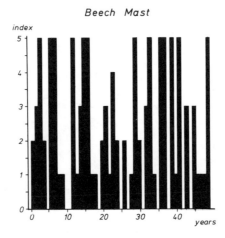

Fig. 7:   Sequence of winter temperature and beech mast indices generated by a computer program for a time period of 50 years.

under exactly identical environmental conditions over a long time period and look for the percentage of populations becoming extinct. This is impossible!

We can , however, simulate such a study with the help of computer programs or analytical methods. First, realistic series of environmental effects are produced (fig. 7). We can simulate the reaction of a population to environmental impact with the help of a realistic population model which considers extensive ecological and demographic data of the population concerned (fig. 8).

step 4

This simulation has to be repeated at least 500 times. The procedure is stopped as soon as "the computer population" has become extinct or a certain time period is over (e.g. 100 years) (fig. 9). Even under constant environmental conditions the population development is not entirely predictable. Stochastic demographic events, if they are simulated in our program, will also have different consequences under absolutely identical environmental conditions.

As a rule, we want to predict the **future chances** of a population to survive either under constant current or modified conditions as projected in regional landscape planning. We know about the probability of certain environmental events but due to their stochastic nature the future cannot be predicted exactly. Therefore, we will not use constant environmental conditions for each of the "simulation populations", but conditions which are generated anew in each simulation according to the known regularities.

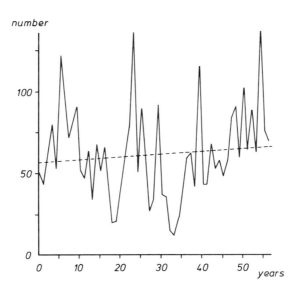

Fig. 8:   Development of population size of the simulated nuthatch population in response to the events generated in Figure 7.

step 5

The results of the simulation can be summarized in a cumulative curve of survival (fig. 10). From this curve we can see with which probability a population will survive for a certain period of time under certain conditions. We can modify these conditions (e.g. larger area size, increasing isolation), again carrying out 500 simulations. **In this way we can**

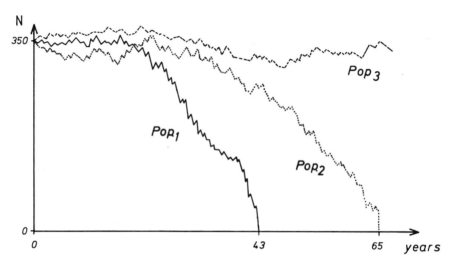

Fig. 9:   Three examples of simulated development of a nuthatch population. In two out of three cases the populations go extinct within the 100-year period.

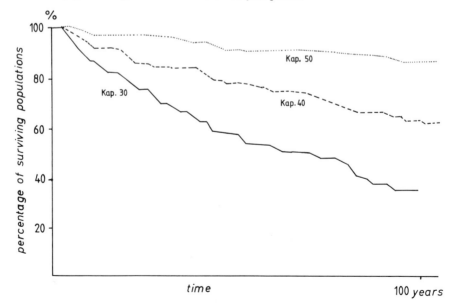

Fig. 10: Survivorship curves after 500 simulation runs each. Habitat capacities of 30, 40 and 50 individuals have been used for the simulations.

delimit conditions which will allow survival of the population with the selected probability. The population which can persist under these conditions is the minimum viable population which fulfils the specified MVP criterion.

step 6

In many cases, we will find that the populations of endangered species do not correspond to the specified MVP criterion (fig. 11 a). With help of computer simulation, we can try to determine conditions which make continued existence possible. In the simplest case, this can be achieved by enlargement of the area size which increases the mean population size to or above MVP level (fig. 11 b). Perhaps measures can be designed which may reduce the required size of the MVP e.g. by reduction of the population fluctuation or reduction of certain risks (fig. 11 c). Suppose, for example, that for the Californian checkerspot butterfly a nature reserve already exists on a southerly slope the inclusion of a northerly slope into the reserve may result in the same protection effect as the inclusion of another, three times larger southerly slope. The population could move from the moister northerly slope to the drier southerly slope according to the climatical conditions. This

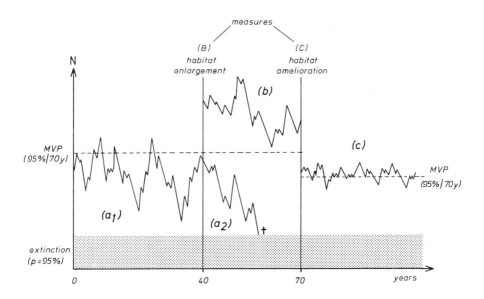

Fig. 11: Fictitious dynamics of an animal population: Within period a_1 the population stays in existence, but during period a_2 it goes extinct. The hatched line indicates the average population size requirement to meet the MVP-criterion defined by a 95% chance to survive the next 70 years. By changing environmental conditions. the population size may be increased (step b), or population dynamics may be reduced (step c). In the first case the average population size is higher than the MVP, while with action c a smaller number of individuals will meet the MVP-criterion.

would reduce the amplitude of population fluctuations and could result in the persistence of a MVP with a lower mean population size than would otherwise be required.

In conclusion: under a specified MVP-criterion there will be not only one but many different conditions which enable the existence of a MVP.

The steps outlined above give the course of an approach to determine a MVP. Especially when formulating a model of population response to several environmental situations a detailed knowledge of the biology of the species concerned is mandatory. Investigating the biology of the species as well as the formulation of a population model are part of the **PVA**. The following information has to be compiled and evaluated (see MARCOT *et al.* 1988). For all data the method used and a confidence interval should be mentioned.

I. State of the population
1.   Recent occurrence and abundance of the target species.
2.   Population development within the last decade. Where did populations disappear? Are the reasons known? Where did new colonizations or re-colonizations take place?
3.   Deterministic reasons for the vulnerability such as destruction of habitats or direct persecution.

II. Area requirements
4.   Social or reproductive units of the species and lowest number of individuals of which a functioning social or reproductive unit consists.
5.   Fluctuation and mean size of home range for the social and reproductive unit, respectively, and their seasonal changes (life range) as well as overlap. Territory size. Dependence of these variables on habitat quality (see points 17-19).
6.   Are food or breeding territories defended? What are the differences in this respect between age, sex and social rank, respectively?

III. Population biology
7.   Absolute population size (if possible for several years).
8.   Maximum possible (short term) population size in the habitat (may exceed long-term carrying capacity).
9.   Population structure with respect to age and sex.
10. Demographic variations in reproductive rate, survival, age of first and last reproduction, dependence of these variables on habitat characteristics (see point 20 and 21) as well as on age and sex.
11. Individual reproductive success (separately for age classes) and their variance (offspring divided into male and female, if possible), necessary for calculation of N_e (see point 13).
12. Mating and breeding system (monogamy, polygamy, philopatry, brood care).
13. Assessment of the effective population size (number of individuals contributing to the gene flow) and estimation of inbreeding coefficients to determine loss of genetic variation.
14. Dispersal behaviour within the habitat or the population (how far away from their parents will offspring breed? Data should be classified according to sex, age, habitat quality, and population density as far as possible).
15. Structure of metapopulation: is the species divided into subpopulations? How much gene flow occurs between subpopulations? Do immigrating individuals contribute to genetic exchange? How great are emigration and immigration rates?
16. Interactions with other species such as competitors, predators, parasites, and animal prey or food plant species.

IV. Habitat quality
17. Primary habitat and substrate for reproduction and raising of offspring.
18. Kind, amount, quality, and structure of microhabitats which promote survival and reproduction .

19.   Habitat qualities used for foraging.
20.   Dynamics of the habitat (are habitats permanently lost e.g. due to succession and when and where do new habitats arise?).

       The **future development** of a population can then be assessed by evaluating points 1-20 as well as considering the following aspects.
21.   Size, number, and spatial relationship of the habitats.
22.   Identification and mapping of suitable habitats.
23.   Dispersal ability and maximum distance between suitable habitats. A circle with half the maximum dispersal distance as radius may be used as a first approximation for practical purposes.
24.   Assessment of the effects of isolation due to habitat fragmentation. Barrier effects of natural and anthropogenic landscape elements. Which structures in the landscape promote exchange of individuals between subpopulations?
25.   (Rare) natural catastrophes, e.g. forest fires, flooding, epidemic events, introduction of alien species, as well as man-made catastrophes (but no permanent changes such as habitat modifications which belong to deterministic threats).

5. The target species concept

5.1. Habitat protection by preservation of target species

To achieve long-term efficiency of nature conservation measures, regular controls are required. Controls are only possible for clearly defined goals.

We think that the goal of long-term species protection is to maintain the *species-specific requirements* in nature. On that basis, we can select "representative species" from various habitats as target species (see MÜHLENBERG 1989). By monitoring populations of such target species, the quality and the success of conservation measures can be assessed and, if necessary, be improved.

The persistence of the species requires a reduction of general adverse environmental influences (e.g. eutrophication, pesticide use, or alteration of water balance). Since an extensive reduction of these adverse factors will not be possible in the near future, we need a continuous monitoring of target species to recognize negative trends and to start suitable counter-measures.

According to area requirements of a long-term viable population, there are different levels of implementation on which effective protection measures can take place (tab. 2):

Other endangered species will also benefit from protective measures for target species: in the case of the Ortolan Bunting (*Emberiza hortulana*), e.g., Lesser Grey Shrike (*Lanius minor*), Woodchat Shrike (*Lanius senator*), Wryneck (*Jynx torquilla*), Little Owl (*Athene noctua*), Garden Dormouse (*Eliomys quercinus*), several butterfly species of orchard trees as well as spiders and beetles of field margins. The protection of target species in hierarchic levels (e.g. with representatives of the various trophic levels in all size classes) guarantees the protection of a whole area including most of the different habitat characteristics.

Tab. 2: Target species with various area requirements have to be managed in different levels of implementation.

Levels of implementation	Target species (example)	Category in German red data book
with farmers	land snail (*Vertico heldii*)	1
single communities	Lesser Purple Emperor (*Apatura ilia*)	3
rural districts	Ortolan Bunting (*Emberiza hortulana*)	1
federal states	Otter (*Lutra lutra*)	1
entire federal territory	Black Stork (*Ciconia nigra*)	1

What we know today is sufficient for immediate protection measures. However, concurrent research and monitoring is absolutely necessary for future improvement and optimization of the measures.

The expenses for the protection measures have to be carried by three ministers. on the federal level, measures for habitat design (including buying of areas) have to be paid for by the Ministry of Environment (BMU). Compensation payments and structural alterations in agricultural and forestry practices have to be financed by BML (Ministry for Agriculture & Forestry; about 75% of the costs). The BML is also responsible for the maintenance of the habitats and continued cultivating practices (about 20% of the needed expenditures); these costs should be kept as low as possible. The BMFT (Ministry for Science and Technology) has to finance accompaning research (about 5% of the costs). The legislation has to provide the adapted legal instrument for effective nature conservation.

We would like to emphasize that the aim of the target species concept is primarily to provide sufficiently large, suitable areas by securing existing habitats (if possible by buying). If there are not enough suitable areas available, then adequate new habitats must be created to the best of our knowledge. Direct supportive measures for the target species (e.g. nest boxes, guarding nests, regulation of competitors or predators) are *not* part of the target species concept. This point clearly separates the target species concept from more traditional ways of species conservation programs.

The concept of protective measures with accompanying research can be successful if carried out for at least two legislative periods (2 x 4 years). A minimum of eight years is needed for the collection of important data concerning degree and causes of yearly or irregular population fluctuations.

We have to accept the fact that these measures and the monitoring will require considerable expenses. As a result, however, we can expect a more effective protection of our natural environment with more transparency in the use of tax money.

5.2. How to select target species

We recommend the following criteria for the selection of **target species** on which nature conservation should concentrate in the future:

1. Species which are the most **endangered nationally or internationally** should enjoy top priority their loss is not reversible. The aim is to reduce the degree of threat (lower rank in the red data list). How much a species is endangered is assessed according to existing populations as well as to the area of distribution (index of chorology after KUDRNA 1986).
2. We want other (endangered) species to benefit from money spent for protective measures. Therefore, especially those species should be selected which are **mainly threatened by alterations in their habitat** and less by direct persecution. Species with a patchy distribution in specific types of habitats are to be preferred to ubiquists. Special attention should be paid to species which are restricted to certain stages of succession.
3. **Keystone species** should be preferred because their extinction would lead to the extinction of many other species.
4. It is easier to manage species which occur only in **our geographical sphere of influence**. For long-distance migrants, **international measures** are needed at the same time.
5. The **chances** of persistence of populations depend on the financial support available if there are other human interests. Here realistic limits have to be considered.
6. The readiness to invest money increases with the **popularity** of the supported target species. Appropriate public relations are always necessary to promote such conservation projects.

The selection of target species results in a priority list for a certain large area, e.g. Middle Europe or Germany, which should be adapted for the purpose of formulating conservation goals on a regional level. The following criteria can be used to set up a *Regionally Modified List of Target Species* (RAZ):

7. Do suitable habitats exist locally?
8. Is the species represented in the region?
9. How large is the local population?
10. Are there realistic prospects for persistence of the species (e.g. under consideration of regional planning?)
11. Large-scale conservation policy by means of the target species concept requires **representative species with various area requirements** to be selected in hierarchical order. Top priority should be given to larger species (large herbivores) with large area requirements (umbrella species, WILCOX 1984) or to those species which are at the top of trophic levels with high metabolic requirements (top mammalian carnivores, birds of prey, and owls). Even when the demands of these umbrella species are satisfied, the requirements of smaller species with special needs in microhabitats may not be fulfilled. Therefore in the same regions, also target species have to be selected which need smaller areas with special habitat characteristics, e.g. old orchard trees, forest edges, wet or dry habitats. Based on local knowledge about distribution and habitat requirements, butterflies, for instance, could be used as "indicator taxa" (WILCOX 1984).
12. In an area without species of supraregional importance, we can select locally endangered species. The protection strategy can thus become effective on a finer scale in small areas.

13. If no possible target species can be found (e.g. ducation (nature trails) or protection of specific ecological functions (e.g. water cycle).

On the whole, all selected target species (for example from the German priority list) should be taken into account somewhere in the locally adapted lists of target species. Large species require coordinated action in several regions.

The **most important aspects of the target species concept** can be summarized as follows:

- the area requirement for protection and management purposes is gauged by the viability of animal populations,

- persistence of populations should be possible under natural conditions,

- information on population biology is needed for the consideration of risk factors,

- quantitative statements for management are possible based on scientific research,

- protection of target species results in protection of habitats, and

- many other species benefit from the protection measures; community ecology can tell us which species these are.

6. Realization of the target species concept

We have seen in chapter 3 and 4 that conservation strategies should turn to population biology as the basis for management. The arguments for and against an approach at the ecosystem or population level are summarized in table 3.

6.1. Evaluation of habitats

To ensure a special conservation status for particularly valuable habitats, "objective criteria for evaluation" have to be developed (SUKOPP 1971). Characteristics of habitats have been searched for which could be indicators of an "intact environment". This problem has not been solved so far and there may be no solution for it at all. Many supposed criteria for intact habitats are taken into consideration and finally no one knows for which populations the evaluation took place. Habitats are too complex to be combined successfully into a linear rank order of improving quality. In our opinion it makes no sense to evaluate different habitats with always the same standardized methods and the criteria discussed above.

Tab. 3:   Arguments of two different approaches to conservation of nature.

Approach at the ecosystem level	Population level
Meet the aim to maintain diversity. An ecosystem supports many species.	it is impossible to carry out a specific management for all species.
Measures can be taken at once.	First of all, a specific knowledge of population biology is necessary.
The approach at the ecosystem level corresponds to the actual scheme of landscape planning in Germany.	The borders of administrative authorities do not match the distribution of populations.
Conflicts hardly arise so long as people want to protect the remnants of nature areas.	A discussion about the selection of target species is essential.
There is a broad consensus in what concerns the priorities of important areas.	Priorities must be read from the RAZ-list (see chap. 5.2).
It is difficult to find out indicator-species for a certain community.	Preference of key species and mobile link species for conservation.
Species lists are used as arguments with no information on the chances for survival of the included species.	Abundances and viability of selected species are used as arguments.
An ecosystem will still remain even if there is some loss of species.	The extinction of a species is absolutely irreversible.
Ecosystems are generally limited by typological aspects of plant associations.	Home ranges of animals do not often cover one special type of plant association.
Definitions of "equilibrium" or "stability" cannot be given exactly Dynamics of the system is usually not taken into account.	The state of a population can be scientifically determined. All the management actions can therefore be controlled precisely.
Minimum area size for ecosystems cannot be assesed. Therefore it is difficult to find objective arguments against the further reduction in size of an ecoytem.	Minimum area requirements are determined by the MVP-concept. The values can be used for specified claims to the politicians.
There are no arguments for a network of habitats and for multiplying reserves to minimize the risk of extinction.	The metapopulation concept calls for the consideration of dispersal abilities. These informations are hardly available.
Conservation is limited on declared areas.	Conservation covers the whole home range of species including farmlands and other cultivated landscapes.

First of all, we have to determine a goal: which characteristics of habitats should be evaluated for what purpose? One possible and legitimate basis for habitat evaluation could be their value for recreation, leisure and education, research, their natural condition, and abundance . CHANTER & OWEN (1976) e.g. developed an index for visitors in nature reserves. Regarding the endangered status of many indigenous species we believe that *available funds should be invested directly in the conservation of endangered species. We should not believe in saving endangered species with "natural", representative, or "scientifically interesting" habitats.* Therefore, we regard the evaluation of habitats under the aspect of preservation of species. The value rises and falls according to the importance of an area for the viability of endangered populations. One or several selected endangered species can indicate intact habitats. We do not have to find out which criteria an area fulfils. This concept of conservation of species is neither new nor the result of scientific research but a consensus in our society and has resulted in laws - perhaps not yet consequently enough.

The following consequences arise for the problem of evaluation:

Suitable criteria for evaluation of habitats may be directly derived from the biology of the selected species. It is a task of the ecological sciences to investigate relations between quality of habitats and the density, and the development, respectively, of populations for particular cases.

E. GLÜCK (pers. comm.) e.g. developed the following method to define the habitat quality: In certain areas, he offers nest boxes for great and blue tits (*Parus major* and *P. caeruleus*) and surveys the successful raising of nestlings. The resource availability is estimated according to a cost-benefit model, in which weight increments of the nestlings are entered as benefits and flight efforts of the adults as costs.

Effective nature conservation cannot mean to rank the last natural areas by decreasing "ecological quality" (e.g. by means of continuously decreasing number of species). This could lead to the transformation of the "less valuable" areas, e.g. for road construction.

It is too defensive as a strategy and - in our view - a waste of energy if nature conservation sees its primary goal in providing planners with a list of areas (ranked according to generally accepted criteria) the loss of which would be least harmful.

6.2. Assessment of environmental impacts

In practice, nature conservation deals mostly with impact assessment in habitats and with concepts for compensation regulation. This requires an **ecological evaluation** of the encroachment.

We are mostly occupied with consequences of impacts on habitat size and fragmentation. Referring to the arguments presented in the last chapter we believe that even when the goal is evaluation of the **habitat change**, the consequences of habitat alterations can best be analyzed quantitatively by evaluation of future chances of **populations affected**.

Firstly we have to establish an **inventory** in the respective habitats. It is more important to obtain reliable species lists of selected taxonomic groups rather than attempting to include as many taxa as possible. The biology of many of these will be widely unknown so they cannot be used for any further assessment anyway.

We need at least one year (all seasons) to establish an inventory.

From these lists, we will select those species which will probably be influenced according to the type of management plan. This step excludes species which are only temporary visitors in the area as well as species which will not be threatened by the special type of alteration (e.g. black woodpecker (*Dryocopus martius*) will probably not be influenced by the canalization of a stream). This list of potentially affected species will be different depending on the different **alternatives in planning**.

The next step divides the list into species which are on the so-called RAZ (**regionally adapted list of target species**, details in chapter 5) and others. The latter are usually not endangered species or species which have no suitable habitat in that area.

For all species, a quick prognosis (**Schnellprognose, SCHNEP**) is carried out. For some species, a clear evaluation of the impact may already be possible.

For non-endangered species, it may be possible to predict where they will disappear and where a population decrease will have to be tolerated. For endangered species statements about isolation effects etc. may be possible.

The resulting prognoses for different species may be compiled for the different planning alternatives.

This may already allow a submission for a judgement by the responsible politicians.

An extensive analysis of the **risk for animal populations** (PVA) will be necessary if SCHNEP does not result in the clear evaluation of consequences of the impact for particular species. This may already have been carried out in the scope of creative species conservation in the region. This is the case when, for instance, the separation of two subpopulations by road construction may eventually lead to the extinction of this species in the whole region. A PVA should always be carried out if SCHNEP indicates clearly lowered viability of a RAZ species.

Again after completing PVAs, a proposal for public decision is possible. This proposal should clearly demonstrate the effects of the different planning alternatives und of the "zero option". With variant A, we expect species 1, 2, 3 to disappear out of a certain area, variant B reduces survival probability of RAZ species 5 in Lower Bavaria from 70% to 30 % during the next 50 years. A weighting of these statements has then to be performed by the political institutions or by legal regulations. The various aspects of process of impact assessment are brought together in a sketch (fig. 12).

Fig. 12: Sketch of an impact assessment

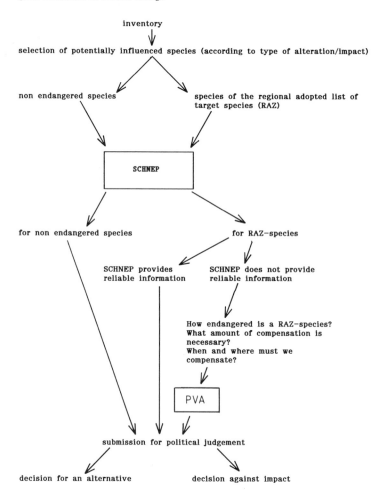

goal: evaluation of habitat change

inventory

selection of potentially influenced species (according to type of alteration/impact)

non endangered species

species of the regional adopted list of
target species (RAZ)

SCHNEP

for non endangered species

for RAZ–species

SCHNEP provides
reliable information

SCHNEP does not provide
reliable information

How endangered is a RAZ–species?
What amount of compensation is
necessary?
When and where must we
compensate?

PVA

submission for political judgement

decision for an alternative

decision against impact

Only in case of dubious prognosis is a thorough PVA to be performed. Should a creative conservation concept (see 6.3) already exist for the region, any impact assessment must be considered in light of the goals put forward by the concept.

6.3. Implementation of creative conservation programs

After the creation of RAZ, which selects those species suitable as target species for a certain area, the current status of this species needs closer investigation. A first step should be taken in a similar way as proposed above, by performing a SCHNEP. The whole process of evaluation as well as developing and implementing a management plan for the benefit of a target species should be supervised by an advisory board of experts. The task of this board is to coordinate the scientific investigation, contacting those institutions and persons who could be affected by any management actions and submitting any proposals for public judgement. It should set up information programs, and work in close contact with those panels preparing conservation programs for other target species in the same region or for the same species in other regions. It is important to recognize if any conflicting demands exist for the protection of different target species. In such a case priorities should be set up considering the supraregional distribution of the species. The board should include an experienced conservationist, an expert on the biology of the target species and members of the administrative units directly involved in conservation management planning. If immediate actions do not succeed in protecting the target species, a thorough PVA for this species has to be initiated to investigate additional possibilities for the protection of this species. It is of great importance, to contact the public repeatedly during this process. Conservation programs will only be successful, if the program is readily supported by public opinion. One way of implementing creative species conservation is presented in a flow diagram (fig. 13).

Fig. 13: Sketch of an creative conservation program

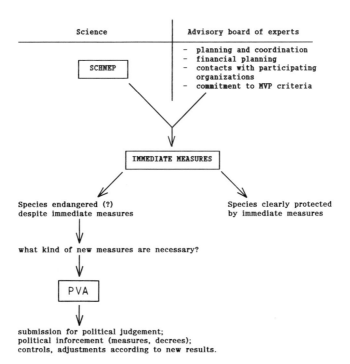

6.4. Performing a quick prognosis (SCHNEP)

The quick prognosis (SCHNEP) is the first important step by scientists for creative species conservation as well as for evaluation of impacts. The goal of SCHNEP is an assessment of the **future chances** of survival of the affected species without carrying out further research to avoid time delays. SCHNEP should not last longer than at **most half a year**. A SCHNEP consists of several steps:

6.4.1 Review of literature

The literature should be consulted in answering the following questions:

- Is there any information available about the most important **reasons for threats**? In many cases the direct threats to a species are certainly known, or well-founded hypotheses may exist.
- Do we know about similar situations of other populations of this species elsewhere (foreign countries)?
- What is known about **habitat size requirements**? How large is the home range of individuals in stable populations? What are the reasons for possible variations of home ranges?
- Are there any studies about correlations between **habitat quality** and **population density**? It would be very helpful to obtain habitat-suitability-index-models (HSI, within the habitat evaluation procedure, HEP; see PEARSALL & DURHAM 1986). In Germany, so far no study with this aim exists. However, in some cases research data probably may be available which can be interpreted in this direction. Habitat requirements and the different factors by which they are influenced are documented at least in part for many indigenous species - even if not in form of HSI-models.
- What information exists about **dispersal ability** and dispersal motivation of individuals of a population? This includes many questions: Which animals leave their home areas when and under what conditions, how far do they migrate, and what are the chances of these "emigrants" reproducing successfully? What landscape structures support (corridors) and prevent (barriers) dispersal?
- Have **fluctuations in density** of populations been monitored over several years? How large are these and what do they depend on? How fast can populations regenerate after having gone through a bottleneck?
- Finally, for the interpretation of rough census data particularly for vertebrates, it is important to know how much smaller the **effective population size** is compared to the size of the total population (see SCHONEWALD-COX *et al.* 1983, HARRIS & ALLENDORF 1989). This is especially necessary for a reasonable prognosis of the genetic risks of a population.

If information on several of these aspects is lacking, one could use data from **ecologically similar** species and make predictions using **allometric equations**. Allometry is used to describe differences of proportions (forms and events) and to correlate them with changes in absolute body mass of the individuum (GOULD 1966).

As an example; due to correlation between metabolism (or energy demands) and body mass, the size of a foraging area is correlated with body mass. PETERS (1983) and CALDER (1984) provide equations on relationships between home range and body mass. These studies also document correlations between mass and abundance of vertebrates - separately for herbivores, carnivores and birds. BELOVSKI & SOULÉ (1987) use these relationships to estimate MVPs of mammals (95 % viability for 100 and for 1000 years).

6.4.2 Gathering preliminary data about the affected population in the field and use of local information

Time is the prime factor determining practical procedures for the collection of data in the field. The following aspects have to be taken into account:

- Do we know the distinct **factors of jeopardizing** the species in the area? A rough survey of the distribution of the threatened species has to be carried out in the whole area. Often, documentations about occurrences particularly of "rare" species may exist locally (mostly unpublished).
- An overview of **habitats potentially suitable** for the species affected is necessary. This should be done with reference to habitat mapping results.
- For the evaluation of the status of the population, it is important to estimate **population size** and **density**. In case no current data are available, rough surveys have to be carried out at the site.
- Equally important is any information about fragmentation of the populations and possible **exchange of individuals**.

It is important to obtain **insights into regional planning** i.e. rural developments and planned impacts before starting the prognosis, which would otherwise be worthless.

6.4.3 Making a prognosis

First, we have to decide whether long-term effects on the population are to be expected which exceed the direct impact. As an example, the blackbird with its well-known contiguous distribution in forests and parks would certainly disappear only in the limited area cleared. We do not expect an effect on its occurrence in neighbouring areas. For such species a detailed estimation of risk is not necessary. It is only the question of locally limited disappearance. This decision and its reasoning is part of the ecological evaluation.

The next step is to determine the unit to which the SCHNEP refers. When the population is completely isolated, the prognosis is restricted to this isolated population. In case of rare exchange of individuals with other subpopulations, the fate of these neighbouring populations has to be included in the prognosis. These considerations are based on the metapopulation concept (SHAFFER 1985). It might be that a road construction isolates subpopulations which thereupon become extinct. In case of frequent exchange with other populations, the whole population is the unit of the prognosis.

For the development of prognoses we have to use criteria adapted to every particular species, e.g.:

-   for large mammals or birds, the effective population size is needed. MARCOT *et al.*
    (1988) provide a table as a rough guideline. For many insects, the population size per
    se should not play an important role for future survival (EHRLICH 1983). Habitat
    dynamics and potential for recolonization are probably more important. Again, a table
    by MARCOT *et al.* could be used as a rough guideline. Another possibility is the use of
    existing population models (e.g. BELOVSKI & SOULÉ 1987, after GOODMAN). Of
    course, the value of prognoses is limited in general models which consider only growth
    rates and variances.

We emphasize that SCHNEP can only be a makeshift. It is always a compromise
between the necessary scientific thoroughness and the time pressure which does not allow
long-term studies. The direct threat to many species by the increasing destruction of their
habitats calls for immediate action. On the basis of SCHNEP immediate measures can take
effect which may have to be modified and extended following more detailed investigations
(PVA).

7. Research requirements and future developments

Population vulnerability analyses (PVAs) of selected animal populations have to be
carried out. Necessary are exemplary studies on target species according to the questions
formulated in chapter 4.4.4. These studies will help us to develop population models for
these target species. They are a necessary prerequisite for the determination of the MVP
(see chapter 4.4.3.) as well as a catalogue of measures which are necessary for the
protection of species and implementation of measures to offset compensation regulations.
As many species as possible should be studied in respect to population dynamics, dispersal
ability, and habitat quality, to name only the most important aspects.

In addition to research, future development has to deal with needs for action, and
models for **practical implementations** have to be developed. The basis for the
environmental impact assessment should be a regionally adapted list of target species which
fulfills similar functions as the state wide habitat survey (e.g. Bavarian Species and Habitat
Protection Programme, ABSP, RIESS 1986).

7.1. Research on the biology of species

Population dynamics. For any prediction on the future of a species, we have to
understand the reasons for population fluctuations. An analysis of the key factors of increase
or decrease of population sizes has to be carried out.

In order to distinguish environmental factors from demographic ones, different local
populations living under similar climatic conditions and belonging to one metapopulation
should be compared for occurrence of synchronous or asynchronous fluctuations. On the

other hand, differential effects of similar external factors on several local populations can be investigated.

Dispersal ability. The dispersal ability of a population determines its success in replacing a locally extinct population or in colonizing reconstituted habitats. We have almost no information on this; even for birds, the most thoroughly investigated group, data are few and often contradictory (e.g. see BAUER 1987, and table 1, chapter 4.3.).

To strengthen the connection between isolated local populations, the concept of "a network of biotopes (habitat-linking-system, MADER 1988)" has been developed in landscape planning. Corridors or "stepping stone habitats" interconnect isolated habitats. Ecotones, like hedges, are certainly valuable in an otherwise cleared landscape, e.g. after the consolidation of farmland.

Little information exists on whether and to what extent species actually enhance their exchange of individuals through these corridors (MÜHLENBERG 1988). Even closely related species seem to differ considerably in the use of these corridors and often have very specific habitat requirements. Therefore general guidelines about interconnecting structures cannot be given.

We need studies on dispersal behaviour especially for endangered species. Are corridors actually used for dispersal or do species reach the habitat islands by direct flight? Are the individuals in the corridors particular dispersal stages or resident corridor inhabitants? How do corridor attributes like width, habitat structure, and gaps influence its function as a means of enhancing dispersal?

Habitat quality. Habitat quality has so far been evaluated mainly according to presence or absence of certain species. For future chances of a population, the effects of habitat characteristics and elements which influence mortality and fitness of individuals are important (cf. risk analysis, chapter 4.1 with a checklist of relevant questions).

If the suitability of habitats for a given population is to be assessed, a study of habitat-suitability-indices is necessary (HSI, cf. Habitat Evaluation Procedure, HEP, VAN HORNE 1983, PEARSALL & DURHAM 1986). We need HSI-models especially for the target species (see chapter 5) from the regionally adapted target species list (RAZ). We have to study the key habitat components relevant for the species and to analyze the quantitative relation between a key factor (variation of the resource) and the density of the species (for examples see SCHRÖDER 1982 and CARREKER 1985).

Since aspects of conservation of nature have to cover the whole landscape (see chapter 2.2.), it becomes increasingly important to study the possibility of restoring spoiled habitats for the use of (target) species (RINGLER 1989). The destruction or alteration of habitats by man is certainly the most important reason for disappearance of species in Central Europe. To obtain more areas for nature conservation, we have to improve degraded habitats.

Ecological studies of restoration will become increasingly important, particularly in view of the closure of farmlands, in reducing the surplus agricultural production in the EC. The most important questions are: with what measures can we improve which degraded habitats for the benefit of target species for conservation?

Concerning the questions of minimum area size, the **edge effects** become more important with decreasing size of areas. Formerly, edge effect was seen as a rather positive factor which enhances diversity (see e.g. ODUM 1983). However, we must consider some negative consequences (see chapter 3.3.).

So far, only limited attention has been paid to the reactions of animals to **disturbances** from habitat margins. In addition to the isolation effect of a road (see MADER 1981b), the avoidance behaviour of the animals has to be considered. Hunted species (see chapter 3.3., THIOLLAY 1989) and carnivorous animals are especially sensitive so that top carnivores (which are our most endangered species) are most strongly affected. Therefore, we need studies done in the densely inhabited landscape of Central Europe on the correlation between disturbance and the disappearance of species despite the availability of resources (SCHNEIDER 1986, FRENZEL & SCHNEIDER 1987, PUTZER 1989).

The evaluation of disturbance is not only important for the planning of nature reserves with undisturbed zones and their use for recreational purposes. It also has to be taken into consideration for impact assessments. In road construction projects compensation should be a matter of course not only for the road itself but also for the disturbed strip along both sides as well as for the built up areas. The reduction in habitat quality (of these areas) for target species will depend on habitat type and the species itself. We need scientific data to determine the width of such strips (ZANDE *et al.* 1980). According to the Bavarian Nature Conservation law 6 d § 2, a disturbed strip of 100 m width along both sides of the road is seen as deteriorated habitat, e.g. in meadow bird breeding areas   (WERRES 1989, unpublished manuscript).

7.2. Practical implementation for future development

Research results are useless unless they are put into action for the preservation of species. **Regionally adapted target species lists** (RAZ, see chapter 5) and **population vulnerability analyses** (PVA) are prerequisites to successful nature conservation planning.

The already existing species lists from different habitats could be interpreted according to the occurrence of particular species in habitat islands of different sizes. In this way, we could learn more about the minimum size above which habitats are accepted by different species. Such data from **incidence curves** (see chapter 4.) can be of use to some extent in nature conservation planning.

In Germany, we are still not used to dealing with target species and to concentrating all protection needs on the scientifically calculated survival probability of a target species. A well-known example of successful protection efforts is the Spotted Owl *(Strix occidentalis)* in the USA (SIMBERLOFF 1987, LANDE 1988). While ensuring the survival of a viable metapopulation, it was possible to protect a forest area of 13.000 km². Now we have to carry out **population vulnerabilities analyses** for endangered species in Germany and to learn which amount of areas will be really necessary for viable populations. We may infer from these data that the preservation of many species is only possible on a large scale including our neighbouring countries.

Zusammenfassung

Der Flächenbedarf einer Tierpopulation wird bestimmt durch (A) den Raumbedarf der Reproduktionseinheit und (B) der Größe einer überlebensfähigen Population. (A) variiert durch die individuell unterschiedlichen und die im Jahresverlauf schwankenden Aktionsraumgrößen und wird zusätzlich erheblich von der Habitatqualität beeinflußt. Die Überlebensfähigkeit (B) einer Population ist neben deterministischen Faktoren auch von Zufallsprozessen abhängig und daher nur mit einer gewissen Wahrscheinlichkeit abschätzbar. Vier verschiedene (nicht anthropogene Faktoren können selbst in einem geeigneten Habitat zum Aussterben von Populationen führen: demographische und genetische Zufallsprozesse, Umweltschwankungen und Naturkatastrophen. Eine Absicherung gegen Risikofaktoren wird durch Vergrößerung der Population, Erhöhung der Zahl geeigneter Habitate und Verringerung der Isolierung zwischen den bewohnten Flächen erreicht.

Zur Bestimmung der Mindestgröße einer überlebensfähigen Population (Minimum Viable Population, MVP) dient als wichtigste Datenbasis eine sog. Gefährdungsgradanalyse (Population Vulnerability Analysis, PVA). Über sie werden Informationen über die spezifische Populationsbiologie, die für die Zukunftssicherung der Population notwendige Habitatqualität, Flächengröße, Lage und zukünftige Entwicklung der Habitate zusammengetragen. Die Zielsetzung einer MVP (z. B. mit 95%iger Wahrscheinlichkeit die nächsten 100 Jahre überlebensfähig) bestimmt dann die notwendigen Rahmenbedingungen. Für besondere Anforderungen in der Praxis wurde eine sog. Schnellprognose entwickelt.

Sowohl für konstruktive Artenschutzmaßnahmen als auch für Umweltverträglichkeitsprüfungen sollte eine Zielart ausgewählt werden, damit die Flächensicherung eindeutig quantitativ begründet werden kann. Für die Auswahl einer Zielart wurden Kriterien aufgestellt, die regional nach den bestehenden Voraussetzungen angepaßt werden. Mit dem Zielartenkonzept lassen sich auch Habitatbewertungen, ein notwendiger Schritt auch bei Eingriffsplanungen, auf wissenschaftlicher Basis quantifizieren.

An ausgewählten Arten zeigt die Analyse der Daten, daß der Flächenbedarf v. a. wegen der unterschiedlichen Habitatqualität sehr variabel ist und für eine MVP viel höher liegt als bisher eingeschätzt wurde (z. B. für eine Singvogelart schon weit über 10 km^2). Ein allgemein verbindlicher Katalog über den Flächenbedarf einzelner Arten läßt sich nicht aufstellen.

References

ANDREN, H. & ANGELSTAMM, P. (1988): Elevated predation rates as an edge effect in habitat islands: experimental evidence. Ecology 69, 544-547.

ANDREWARTHA, H.G. & BIRCH, L.C. (1986): The ecological web. The University of Chicago Press, Chicago, London, 506.

BAUER, H.-G. (1987): Geburtsortstreue und Streuungsverhalten junger Singvögel. Die Vogelwarte 34, 15-32.

BEGON,M., HARPER, J.L. & TOWNSEND, C.R. (1986): Ecology. Blackwell Scientific Publications, Oxford, London, Edingburgh, Boston, Palo Alto, Melbourne.

BELOVSKY, G.E. & SOULE, M.E.: (1987): Extinction models and mammalian persistence. In: SOULÉ, M.E., Viable populations for conservation. Cambridge University Press, Cambridge, 35-57.

BLAKE, J.G. & KARR, J.R. (1987): Breeding birds of isolated woodlots: Area and habitat relationships. Ecology 68 (6), 1724-1734.

BOECKLEN, W.J. & SIMBERLOFF, D. (1986): Area-based extinction models in conservation. In: Elliott, D.K.: Dynamics of Extinction. John Wiley & Sons, New York, 247-276.

BRITTINGHAM, M.C. & TEMPLE, S (1983): Have cowbirds caused forest songbirds to decline?. Bioschience 33, 31-33.

BROWN, M. & DINSMORE, J.J. (1986): Implications of marsh size and isolation for marsh bird management. J. Wild. Management 50, 392-397.

BURKEY, T.V. (1989): Estinction in nature reserves: the effect of fragmentation and the importance of migration between reserve fragments. Oikos 55, 75-81.

CALDER, W (1984): Size, function, and life history. Harvard Univ. Press, Cambridge, Mass. III.

CARREKER, R.G. (1985): Habitat suitability index models: least tern. Biological Report 82 (10.103)

CHANTER, D.O. & OWEN, D.F. (1976): Nature reserves: a customer satisfaction index. Oikos 27, 165-167.

COLEMAN, B.D., MARES, M, WILLIG, M.R. & HSIEH, Y. (1982): Randomness, area, and species richness. Ecology 63, 1121-1133.

CONNOR, E.F. & MCCOY, E.D. (1979): The statistics and biology of the species-area relationship. The American Naturalist 113, 791-829.

DIAMOND, J.M. & MAY, R.M. (1976): Island biogeography. In: May, R.M.: Theoretical Ecology. Principles and applications. Blackwell scientific Publ., Oxford.

DIAMOND, J.M. (1978): Critical areas for maintaining viable populations of species. In: Holdgate, M. & Woodman, M.J. (eds.): Breaddown and restoration of ecosystems. New York, 27-40.

EHRLICH, P.R. (1983): Genetics and the extinction of butterfly populations. In: Schonewald-Cox, C.M., Chambers, S:M., Mac Bryde, B., Thomas,L.: Genetics and conservation. Menlo Park/CA Benjamin/Cummings, 152-163.

ELLENBERG, H., RÜGER, A. & VAUK, G. (1989): Eutrophierung - das gravierendste Problem im Naturschutz?. Norddeutsche Naturschutzakademie, 2.Jahrgang., Heft 1, Schneverdingen, 70.

ERZ, W. 1981: Flächensicherung für den Artenschutz - Grundbegriffe und Einführung. Jb. Naturschutz Landschaftspfl. ABN 31, 7-20.

FRANKEL, O.H. & SOULÉ, M.E. (1981): Conservation and evolution. Cambridge University Press, Cambridge.

FRENZEL, P. & SCHNEIDER, M. (1987): Ökologische Untersuchungen an überwinternden Wasservögeln im Ermatinger Becken (Bodensee): Die Auswirkungen von Jagd, Schiffahrt und Freizeitaktivitäten. Orn. Jh. Bad.-Württ 3, 53-79.

GOODMAN, D. (1975): The theory of diversity-stability relationship in ecology. The quaterly review of biology 50, 237-365.

GOODMAN, D. (1987): The demography of chance extinction. In: Soule, M.: Viable populations for conservation.. Cambridge University Press, Cambridge, 11-34.

GOULD, S.J. (1966): Allometry and size in ontogeny and phylogeny. Biological Reviews of the Cambridge Philosophical Society 41, 587-640.

HAARMANN, K. (1979): Sind Naturschutzgebiete für die Erhaltung in der BRD gefährdeter Brutvogelarten geeignet?. Vogelwelt 100, 70-77.

HAARMANN, K. (1983): Der aktuelle Zustand der Naturschutzgebiete in der BRD - eine vorläufige Übersicht. Schr. R. Drl. 41, 27-31.

HAMPICKE, U. (1988): Extensivierung der Landwirtschaft für den Naturschutz - Ziele, Rahmenbedingungen und Maßnahmen. Schriftenreihe Bayer. Landesamt für Umweltschutz, Heft 84, 9-35.

HARRIS, R.B. & ALLENDORF,F.W. (1989): Genetically effective population size of large mammals: an assessment of estimators. Conservation Biology, 3, 181-191.

JANZEN, D.H. (1983): No park is an island: increase in interference from outside as park size decreases. Oikos 41, 402-410.

JENNERSTEN, O. (1988): Pollination in *Dianthus deltoides* (Caryophyllaceae): Effects of habitat fragmentation on visitation and seed set. Conservation Biology 2, 359-366.

KANGAS, P. (1987): On the use of species area curves to predict extinctions. Ecol. Joaety Bull 86, 158-162.

KUDRNA, K. (1986): Grundlagen zu einem Artenschutzprogramm für die Tagschmetterlingsfauna in Bayern und Analyse der Schutzproblematik in der Bundesrepublik Deutschland. Nachr. Ent. Ver. Apollo, Frankfurt, 6, 1-90.

LANDE, R. (1988): Demographic models of the northern spotted owl (Striex occidentalis caurina). Oecologia 75, 601-607.

LEHMKUHL, J.F. (1984): Determining size and dispersion of minimum viable populations for land management planning and species conservation. Environmental Management 8, 167-176.

MACARTHUR, R.H. & WILSON, E.O. (1963): An equilibrium theory of insular zoogeography. Evolution 17, 373-387.

MADER, H.-J. & MÜHLENBERG, M. (1981): Artenzusammensetzung und Ressourcenangebot einer kleinflächigen Habitatinsel, untersucht am Beispiel der Carabidenfauna. Pedobiologia 21, 46-59.

MADER, H.-J. (1979): Die Isolationswirkung von Verkehrsstraßen auf Tierpopulationen untersucht am Beispiel von Arthropoden und Kleinsäugern der Waldbiozönose. Schr.-R.-Landschaftspfl. Naturschutz H. 19, 131.

MADER, H.J. (1981): a: Untersuchungen zum Einfluß der Flächengröße von Inselbiotopen auf deren Funktion als Trittstein oder Refugium. Natur und Landschaft 56, 235- 242.

MADER, H.-J. (1981): b: Der Konflikt Straße - Tierwelt aus ökologischer Sicht. Schriftenreihe für Landschaftspflege und Naturschutz 22, Bonn-Bad Godesberg, 104.

MADER, H.J. (1983): Warum haben kleine Inselbiotope hohe Artenzahlen?. Natur und Landschaft 58, 367-370.

MADER, H.-J. (1985): Welche Bedeutung hat die Vernetzung für den Artenschutz?. Schriftenreihe Deutscher Rat für Landespflege 46, 631-634.

MADER, H.-J. (1988): In: Schreiber, K.-F. (eds.): Connectivity in landscape ecology. Münstersche Geographische Arbeiten 29, Münster 1988, 97-100.

MARCOT, B.C., HOLTHAUSEN, R. & SALWASSER, H. (1988): An Assessment Framework for Planning for Viable Populations. Manuscript, 43.

MARTIN, T.E. (1988): Habitat and area effects on forest bird assemblages: Is nest predation an influence?. Ecology 69, 74- 84.

MAYFIELD, H.F. (1977): Brown-headed cowbird: Agent of extermination?. Am. Birds. 31, 107-113.

MÜHLENBERG, M. & WERRES, W.: (1983): Lebensraumverkleinerung und ihre Folgen für einzelne Tiergemeinschaften. Experimentelle Untersuchungen auf einer Wiese. Natur und Landschaft 58, 43-50.

MÜHLENBERG, M. (1988): Konzeptentwicklung und Möglichkeiten praktischer Umsetzung von Biotopverbundsystemen. In: Biotopvernetzung in der Kulturlandschaft II, Schriftenreihe Angewandter Naturschutz, Lich, 5, 14-31.

MÜHLENBERG, M. (1989): Freilandökologie. 2. Auflage, UTB, Quelle & Meyer, Heidelberg, Wiesbaden,432.

MÜHLENBERG, M. (1990): Langzeitbeobachtung für Naturschutz - Faunistische Erhebungs- und Bewertungsverfahren. Ber. ANL, 14, 79-100.

ODUM, E.P. (1983): Grundlagen der Ökologie. Band 1, Georg Thieme Verlag Stuttgart, New York, 476.

OPDAM, J & SCHOTMAN, A. (1987): Small woods in rural landscapes as habitat islands for woodland birds. Acta Oecologica/Oecologica Gener. 8 (2), 269-274.

PEARSALL,S.H., DURHAM, D. & EAGAR, D.C. (1986): Evaluation methods in the United States. In: Usher, M.B.: Wildlife conservation evaluation. Chapman and Hall, London, New York, 111- 133.

PETERS, R.H. (1983): The ecological implications of body size. Cambridge University Press, Cambridge, 330.

PICTON, H.D. (1979): The application of insular biogeographic theory to the conservation of large mammals in the northern rocky mountains. Biol Conserv. 15, 73-79.

PLACHTER, H. (1984): Zur Bedeutung der bayerischen Naturschutzgebiete für den zoologischen Artenschutz. Ber. ANL 8, 63-78.

PLACHTER, H. (1985): Schutz der Fauna durch Flächensicherung - Stand, Möglichkeiten und Grenzen.Schriftenreihe des Deutshcen Rates für Landespflege 46, 618-630.

PUTZER, D. (1989): Wirkung und Wichtung menschlicher Anwesenheit und Störung am Beispiel bestandsbedrohter, an Feuchtgebiete gebundener Vogelarten. Schr.-R. Landschaftspflege u. Naturschutz 29, 169-194.

RANNEY, J.W., BRUNER, M.C. & LEVENSON, J.B. (1981): The importance of edge in the structure and dynamics of forest islands. In: Burgess, R.L. & Sharpe, D.M.: Forest island dynamics in man-dominated-landscapes. Springer-Verlag, New York, Heidelberg, Berlin, 67-95.

REICHOLF, J. (1980): Die Arten-Areal-Kurve bei Vögeln in Mitteleuropa. Anz. orn. Ges. Bayern 1/2, 19.

REMMERT, H. (1984): Ökologie. Ein Lehrbuch. Springer Verlag, Berlin Heidelberg, New York, 334.

RIESS, W. (1986): ABSP INFO (Informationen zum bayerischen Arten- und Biotopschutzprogramm), 1, 16.

RINGLER, A. (1989): Restitution von Eingriffsflächen und ihre Sukzessionsgeschwindigkeiten. Vortrag am Laufener Ökologie Symposium "Zeit" als ökologischer Faktor, Juni 1989, Ingolstadt.

ROBBINS, C.S. (1979): Effect of forest fragmentation on bird populations. In: Management of southern forests for nongame birds. General Technical Report SE-14. USDA Forest Service. Southeastern Forest Experiment Station, Asheville, North Carolina, 198-212.

SALWASSER, H., MEALY, S.P. & JOHNSON, K. (1984): Wildlife population viability: a question of risk. American Wildlife 49, 421-439.

SCHNEIDER, M. (1986): Auswirkungen eines Jagdschongebietes auf die Wasservögel im Ermatinger Becken (Bodensee). Orn. Jh. Bad.-Württ. 2, 1-46.

SCHONEWALD-COX, C.M., CHAMBERS, S.M. MC BRYDE, B. & THOMAS, L. (1983): Genetics and conservation. Menlo Park, Benjamin/Cummings.

SCHRÖDER, R.L. (1982): Habitat suitability index models: Pine Warbler. FWS/OBS - 82/10.28. U.S. Department of Interior, Fish and Wildlife Service, Office of Biological Services, Washington, D.C., 8.

SHAFFER, M.L. (1981): Minimum population sizes for species conservation. Bio science 31, 131-134.

SHAFFER, M.L. (1985): The metapopulation and species conservation: the special case of the northern spotted owl. In: Gutiérrez,R.L. & Carey,A.B.: Ecology and management of the spotted owl in the pacific northwest. Portland, OR, U.S.D Forest Service, 86-99.

SIMBERLOFF, D. (1987): The spotted owl fracas: mixing academic, applied and political ecology. Ecology 68, 766-772.

SOULÉ, M.E. (1983): What do we really know about extinction? In: Schonewald-Cox, C.M. Chambers, S.M., MacBryde, B. & Thomas,L.: Genetics and conservation, Menlo Park/CA, Benjamin/Cummings, 11-124.

SOULÉ, M.E. (1986): Conservation Biology. Sinauer Associates Inc., Sunderland, Massachusetts, 586.

SUKOPP, H. (1971): Bewertung und Auswahl von Naturschutzgebieten. Schriftenreihe für Landschaftspflege und Naturschutz, Bonn - Bad Godesberg 6, .

TEMPLE, S & CARY, J.R. (1988): Modeling dynamics of habitat interior bird populations in fragmented landscapes. Conservation Biology (4), 340-347.

TERBORGH, J.W. (1974): Preservation of natural diversity: The problem of extinction-prone species. BioScience 24, 715-722.

THIOLLAY, J.M. (1989): Area requirements for the conservation of rain forest raptors and game birds in French Guiana. Conservation Biology 3 (2), 128-137.

USHER M.B. (1980): An assessment of conservation values within a large site of special scientific interest in North Yorkshire. Fld. Stud. 5, 323-348.

USHER, M.B. (1989): Scientific aspects of nature conservation in the United Kingdom. Journal of Applied Ecology 26, 813-824.

VAN DORP, K. & OPDAM, P.F.M., (1987): Effects of patch size, isolation and regional abundance on forest bird communities. Landscape Ecology 1, 59-73.

VAN HORNE, B. (1983): Density as a misleading indicator of habitat quality. Journal of Wildlife Management 47, 893-901.

WEISS, S.B., MURPHEY, D.D. & WHITE, R.R. (1988): Sun, slope, and butterflies: topographic determinants of habitat quality for *Euphydryas editha*. Ecology 69(5), 1486-1496.

WERRES, W. (1989): Naturschutzfachliche Anforderungen an die Eingriffsplanung des Straßenbaus. Teil 1: Naturhaushalt. Manuskript, unveröffentlicht, 18.

WHITCOMB, R.F., ROBBINS, C.S., LYNCH, J.F., WHITCOMB, B.L., KLIMKIEWICZ, M.K. & BYSTRAK, D. (1981): Effects of forest fragmentation on avifauna of the eastern deciduous forest. In: Burgess, R.L. & Sharpe, D.M.: Forest island dynamics in man-dominated-landscapes. Springer-Verlag, New York, Heidelberg, Berlin, 125-205.

WILCOVE, D.S. (1985): Nest predation in forest tracts and the decline of migratory songbirds. Ecology 66, 1211-1214.

WILCOVE, D.S., MCLELLAN, C.H. & DOBSON, A.P. (1986): Habitat fragmentation in the temperate zone. In: Soulé, M.E.: Conservation Biology. Sinauer Associates Inc., Sunderland, Massachusetts, 237-256.

WILCOX, B & MURPHY, D.D. (1985): Conservation strategy: the effects of fragmentation on extinction. Am. Nat. 125, 879-887.

WILCOX, B (1984): In situ conservation of genetic resources: Determinants of minimum area requirements. In: McNeely, J & Miller, K.R. (eds.): National parks, conservation, and development, IUCN and Smithsonian Institution Press, Washington, 639-647.

ZANDE, A.N. van der KEURS, W.J. ter & WEIJDEN, W.J. van der (1980): The impact of roads on the densities of four bird species in an open field habitat - Evidence of a long-distance effect. Biol. Conser. 18, 299-321.

Species Conservation: A Population-Biological Approach
A. SEITZ & V. LOESCHCKE (eds.) © 1991 Birkhäuser Verlag, Basel

The Isolation of Animal and Plant Populations: Aspects for a European Nature Conservation Strategy

H-J. Mader, Rübhausener Str. 42, D-5330 Königswinter 21, Germany

Abstract

Growing efforts for effective and longlasting nature conservation cannot keep pace with the amount and degree of recource utilization and destruction everywhere in the world. Plant and animal species, their communities and habitats are lost every day. Only large scale conservation measures, which are supported by many nations will eventually be able to decelerate this process.

Isolation and fragmentation are among the basic causes for the impoverishment of nature. Minimum area concepts may be misused to restrict and confine nature conservation within the limits of such area definitions.

The European Community is preparing a directive to protect the natural habitats including the wild flora and fauna, aiming at a coherent system of habitats throughout Europe called NATURA 2000. This ambitious directive requires considerable area contribution from the member states, emphasising the protection of the rare and most endangered habitats.

Introduction

Nature conservation of the past decades has failed. There is no doubt that a comprehensive protection of species or habitats has not been achieved, in spite of great efforts and considerable financial input. The causes for this are not restricted to competing agricultural or industrial use of the landscape nor to limited allocation of personal and funds, they may as well be seen in the often strained relations between nature conservation and natural sciences. Nature conservation reacts reserved and sceptical about the growing knowledge of biological and ecological sciences. Vice versa the scientists frequently hesitate to translate their results and knowledge into nature conservation politics. Nature conservation has to correct its image as a nice and fancy idea for wealthy people with a lot of free time by demonstrating that it is an ubiquitous necessity in all political fields and essential for the survival of future generations.

An important step towards this goal is the appreciation of the term population in nature conservation. The population is the most important operational unit in biology. The extermination of species is executed by the stepwise loss of local populations. Hence, species have to be protected by measures which keep the complex of various populations within their habitats vital and capable of development. Species will not be preserved by sectoral efforts for a group of individuals nor by concentrating on a fragmented and isolated small population. Thus, spatial and biological connectivity have to be preserved and developed, and human activities which impair these have to be restricted. It is evident that political frontiers are of little significance since animals and plants do not stop at border-crossings.

TABLE 1:

AREA EFFECTS

*   natural and semi-natural habitats diminish in size

*   natural and semi-natural habitats vanish

*   distances between the remaining habitats grow

*   adverse biotic and abiotic conditions of the matrix   surrounding such habitats are increasing

BARRIER EFFECTS

*   network of linear infrastucture grows larger(roads, forest- and field-paths, railroad-tracks,channels)

*   quality of the network changes (increase in road width, traffic frequency, percentage of paved agricultural roads)

EDGE EFFECTS

*   habitat boundaries become sharp

*   ecotones vanish

*   transitional zones between habitats dislocate from outside to the inside of natural or semi-natural habitats

Insularization and Isolation

In Europe and elsewhere in the industrialized parts of the world, the current trend in landscape ecology, especially in agricultural areas, is still characterized by a reduction and isolation of natural and semi-natural areas, a fragmentation of habitats, and a subdivision of natural populations of plants and animals in small and separated units (MADER 1980, MADER 1984, QUINN & HARRISON 1988). This process is obvious and well documented in literature.

The stepwise but thorough changing of the landscape underlies certain regularities which have to be studied before corrective measures can be suggested and applied. Three groups of effects can be distinguished (Tab. 1)

Most of these effects have been described in comparative studies of maps and aerial photographs and are thus well documented (CURTIS 1956, EWALD 1978, BORCHERT 1981, KNAUER 1985, among others). It can be shown that most of these effects are neither restricted to Central European industrial areas nor to the so- called developed world but that are dealing with a world-wide process of growing ecological importance.

3. Effects of Isolation on Animal Communities

Numerous studies have been conducted on true islands and on so called "habitat islands" with regard to the theory of island biogeography (MACARTHUR & WILSON 1963). In spite of the important impulses which have come from that thinking up to now, the theory is heavily debated regarding its possible applications in nature conservation (DIAMOND 1975, SIMBERLOFF & ABELE 1976, MCCOY 1983, LAHTI & RANTA 1985, MURPHY & WILCOX 1986, SIMBERLOFF 1986 among others). As a result from these discussions, the SLOSS - alternative is raised as an issue of controversy in a non serviceable polarisation (**SLOSS** stands for Single Large Or Several Small areas to be protected). In Europe this quarrel has turned into an even further-reaching conflict between two strategies of nature conservation: the **segregation-** and the **integration-model**.

In the segregation-model, a restricted number of selected and especially valuable areas are effectively protected as nature reserves (sacrificed for nature conservation), while all other areas continue to be used for the different land-use types as before in an unchanged manner, possibly even with growing intensity. Nature conservation politics do not touch or cover these areas.

The integration model on the other hand seeks to cover all areas, looking for possibilities to lower the intensity of land use on a large scale. It tries to combine this with a so called habitat-linking-system (MADER 1988). A habitat-linking-system is built up by a network of unused or extensively used strips of land along waterways, parallel to field-paths, at forest edges, etc. It does not necessarily mean that all leftover natural and seminatural habitats have to be connected physically but the chances for animals to reach these habitats should considerably improve. Considering the presentlimited spatial reserves in Central Europe, only a compromise - i. e. a combination of both strategies as far as possible -makes sense.

The problems and outcome of isolation and insularization can be made clear by the following examples:

1)   Information on endangered species has been compiled on a worldwide scale. On the basis of these data, it can be shown that out of the 77 mammal and bird species having become extinct in the past, 58 (= 75 %) were island species (FRANKEL & SOULÉ 1981).

2)   Similar results are evident from habitat islands. One striking example is reported by DIAMOND, BISHOP & BALEN (1987): At Bogor Botanical Garden (West-Java), the tropical forest surrounding the park was cleared while the habitat structure within the 86-ha park was strictly preserved. Out of 62 bird species which have been counted 1952 (before the deforestation adjacent to the botanical garden started), 20 species have disappeared, four will become extinct soon, and five are suffering from severe decline in abundance.

A series of different factors seem to play a role in structuring species communities and in determining the dynamics of populations in habitat islands. They are not restricted to genetic factors (Tab. 2).

TABLE 2:
___

*   limited resources

*   low habitat diversity

*   reduced number of species

*   attractivity resulting in high numbers of extraneous species and high species turnover

*   sensitivity to disturbance and anthropogenic impacts

*   low buffering potential

*   specialists are replaced by generalists

___

Criteria for a European Initiative on Nature Conservation

The analysis of possible risks from current and expected landscape development for the survival of plant and animal species leads to a set of criteria to be considered when establishing a new conservation strategy. Four criteria of utmost importance are:

Large size

The conservation of small and isolated patches of landscape - allthough valuable and extremely rare habitats may be among them - is insufficient if they are embedded in a matrix of intensively used areas. Gene-flow, immigration, and dispersal must be guaranteed at the very least. Small nature conservation sites suffer from disturbances and many if not the majority of the ecological processes are dominated by external factors from neighboring systems. Even though numerous studies suggest that several small habitats hold more species than one large habitat of comparable size (QUINN & HARRISON 1988), such studies usually do not cover the long term chances for survival of the species nor of the habitats themselves.

Long-lasting protection

It takes decades or hundreds of years for ecosystems to pass through successive stages. Habitats cannot be destroyed and rebuilt like houses or gardens. A mobile and dynamic society must learn and accept that nature's pace is not optional nor can it be manipulated. Hence, nature protection areas must be given a long-term protection status.

Coherence

Conservation sites need biological connectivity. This need not be a physically closed network but could be achieved through buffer-zones, strips of extensively used land, and "stepping stones" which help to increase the heterogeneity of the landscape (NOWAK 1986, MADER 1988, JEDICKE 1990).

Selection of sites

This seems to be the most critical and disputed point. Two aspects should be considered when selecting sites for nature conservation in developed and industrialized regions like Central Europe:

*   first: the **degree of vulnerability**, covering the distribution, the abundance, the dynamics, and the uniqueness of habitats. Endemic species and those habitats which are in danger of extinction should have the highest priority.

*   second: the remaining **large, natural or semi-natural habitats**, even though no spectacular or rare species are found there, are of the same importance. Their contribution to an overall ecological stabilization and to the continuity of evolutionary processes has long been underestimated. On this point, environmentalists and conservationists agree.

These thoughts have to be incorporated into the evolving European conservation strategy, the aim of which has been formulated as follows: the preservation of the common European natural heritage in all its biotic diversity and with a long-lasting perspective. This should be achieved by building a coherent system of special protected areas spread all over Europe.

The Habitat Directive of the European Community

Several attempts have been made to improve the survival chances of rare species within Europe. The well known Bern Convention has issued a comprehensive list of plant and animal species which need protection against direct destruction and whose habitats should be protected. Unfortunately, the Convention never developed the political pressure needed to reach these goals. On the EC-level, the bird-directive was a new start. But its implementation by the member states was reluctant and inadequate, and of course the bird-directive was restricted to birds.

With view to the common market in 1993 and in awareness of the fact that all animal and plant species having the center of their distribution within Europe, deserve - as a common European heritage - a special protection effort, the Commission of the European Community has proposed a directive to **protect the natural and semi-natural habitats including the wild flora and fauna.** As a start, the directive will try to create a system of protected areas on the territories of the member states. These areas should be combined together with the protection areas of the **bird directive** and (possibly) the wetlands nominated for the **Ramsar convention** to form the so called **"NATURA 2000"** system. It is obvious that this system has to be extended as soon as possible to include the non-member states in Europe, especially those in alpine regions and those of Eastern Europe.

To facilitate the selection of the special protection areas, lists of animals, plants, and habitats have been prepared. These include endangered, rare, and endemic species, as well as habitats in Europe that run the risk of disappearing. For each species a certain number or a percentage of its remaining habitats should be preserved and managed according to special plans; if possible, they should be rebuilt and regenerated. There are a series of difficult problems to be solved. How do we use these lists to establish a coherent system of special protected areas of sufficient overall size and connectedness? What method should we use to select and nominate the special protected areas, the necessary size for each habitat or the total area-contribution of each member state? Various models are being negotiated:

-   One model comprises a two stage protection system, assigning, in its first stage, a very stringend protection status (TABU-zones) to the 10 most important habitats of each of the endangered or endemic species and of a restricted number of rare ecosystem-types. In a second step, the 100 most important habitats per species and per ecosystem-type should be incorporated into the system as special protected areas with lower protection status. This proposal was originally made by the Commission. It has earned only very limited support by the member states, mainly because of the arbitrarity of the figures 10 and 100.

-   In a second model a certain minimum area for each species and habitat is calculated within each region of the European Community. The member states are to assign the long-term protection status to those areas. - Another possible solution would be to nominate areas within each region of the Community according to a series of ecological criteria. Such criteria include: abundance and dynamics of species, their distribution, rarity, uniqueness, biological diversity, etc. The use of an algorithm has been proposed to calculate an index of vulnerability which then has to be converted into a percentage of area to be protected.

-   Yet another approach would require each region within the community to contribute a certain percentage of its overall size to the NATURA 2000 system. With regard to the differences in infrastructure (agriculture, percentage of woodland, industrialization), this contribution should allow for variation between regions but should reach a minimum of 10 % of the total area within each member state of the Community. The measures to meet the 10 % requirement and the decisions on areas to be selected should remain the responsibility of the regional authorities - which would rely upon the knowledge and experience of their local experts.

Which model would be the most appropriate in building up the **Network NATURA 2000** is still under debate. Many member states insist upon their exclusive rights to designate habitats as special protected areas and to assign those areas an adequate legal protection status. It seems reasonable to keep this competence at the member state level or (like in the Federal Republic of Germany) at the regional level, rather than transfering it to the European Community authorities. However, there must be an effective tool to correct for possibly hesitatant or restrictive conservation politics in member states, if the ultimate goal - the build-up of a coherent network - is to be achieved.

Within the special protection areas, appropriate measures have to be taken to prevent disturbance of animal and plant communities and to preserve the integrity of the area. Areas have to be developed and managed according to nature conservation goals. An idealistic conception would aim at a system composed of core areas with highest protection status, buffer-zones protecting these areas from external disturbances, and corridors or a "stepping-stone-connection" to insure the mobility and the dispersal of species. In view of the continuously increasing processes of fragmentation of landscapes and of isolation of habitats, it is of utmost importance that this **"NATURA 2000"** system will not be a collection of small and smallest nature oases put together by chance nor a result of political opportunity.

Beside the many technical problems of how to structure and develop the system as a coherent network, and how to ensure that the most valuable and ecologically important areas are selected, there is a severe financial problem to be solved. Many member states insist that the EC should provide for the necessary funds. Only very vague figures have been circulated on the possible costs of this undertaking but it is obvious that a large amount of money will be needed. Therefore, the Commission has recently proposed a new directive (called ACNAT) as a financing tool. However, there are various and grave political objections to be overcome which propably will lead to lengthy negotiations.

Minimum Areas - A Misleading and Dangerous Concept ?

Whatever the political solutions to the problem of nature conservation in different regions of the world may be, there will always remain the questions of how much land must be sacrificed for nature, how densly should the network of protected areas be structured, do we need some large or plentyfull patchy habitats, is there a minimum area requirement for populations which guarantees their survival?

In earlier papers dealing with nature conservation concepts and models, a scientific calculation of **minimum areas for species** or **minimum number of indiviuals for a population** has been required (SHAFFER 1981). It seems that conservationists and landscape planners in their hopeless struggle against land use competitors look for support in undisputable scientific arguments, preferably in the form of thresholds or quantitative (area related) values which can easily be incorporated into their planning. Some biologists have already reacted upon this demand. HEYDEMANN (1981) offered minimum area values for a series of animal species and FRANKLIN (1980) suggested the minimum population size of 500 individuals for mammals. In the last years more sophisticated models have been elaborated providing figures for **minimum viable populations (MVP)** concentrating on **population vulnerability analysis (PVA)** (SOULÉ 1987, HARRIS *et al.*1987, LANDE 1988, MENGES 1990, MURPHY *et al.* 1990). These efforts are necessary and most welcome, although there are quite a fewproblems associated with them. One of the more serious concerns is that the type of information gained from those theories can easily be misused by planners and may lead to a misunderstanding of conservation necessities in the public opinion.

There have been efforts to describe the minimum environment required by a given species for its survival, i.e. its abiotic conditions, nutrition resources, structural demands etc. Government funds have been assigned to such scientific studies and distributed to several ecological institutes in Western Germany. Similar programs are probably on the way in various countries. There are, however, two dangers in such a strategy:

It seems most uncertain that biology will ever be able to supply the necessary data for the calculation of minimum areas (GRUMBINE 1990). Up to now and for a foreseeable future, we are not in a position to gather the basic data on population dynamics for more than a few species. The lack of data on the interactions at higher hierarchical levels - for example at the community level or even at the ecosystem level - is evident. For a restricted number of spectacular species - most of them mammals or birds - we may at present have thorough knowledge about distribution and abundance, for some we may even know the necessary figures to establish a life table. Hence we possess a sufficient basis for the calculation of minimum population sizes or minimum areas for just a very few, selected species. But what about the other 99.5 % for which at present our scientific database is almost negligible?

Furthermore, the calculation of minimum population sizes will end up with totally different figures, depending on whether one aims at a survival for the next ten generations, or whether one wants to include the potential for adequate evolutionary processes, giving such populations chances to adapt to new environmental situations or spatially and temporarily evade catastrophes. Clearly, there is a need for comprehensive definitions of the goals and standards of nature conservation. The point is that the calculations of MVP's will vary with the standards set by society - and if society agrees upon low standards, scientist will calculate the related low figures.

Some or maybe all populations must have the chance to bloom and to build up a surplus of indiviuals from time to time. Such situations may be the starting point for migratory movements or for colonization and the foundation of new populations. Additionally, such productive phases in population dynamics insure the supply of the new alleles. It is important to avoid any limitation to the dynamics of living systems. We have already agreed upon the fact that barriers and isolation will limit the mobility and the areal dynamics of organisms. Minimum population sizes or minimum area concepts for species may hamper the population growth and the areal dynamics of communities if landscape planning is encouraged to implement this concept in a restricted and narrow minded sense.

This leads to the second danger. Politicians and the public will foreseeably react upon this concept in such a way as to limit nature conservation to those established minimum figures. Why should agriculture sacrifice more land than necessary to keep a population of wild plants or animals alive, why spend more money than necessary for a diminuishing species as long as its total number is higher than the scientific calculated minimum population size ? The minimum population argument could in this sense be misused against the efforts to stabilize ecosystems. Biologists and population ecologists may - by following this type of research and by using it as nature conservation argument - help to construct the narrow cage in which the remaining compartments of the ecosystems will gracelessly be enclosed.

It is obvious that research on population ecology and population genetics is of utmost importance for our understanding of living systems. Most questions remain unanswered, and it is the task of the scientific community to fill this gap of knowledge. But nevertheless, we must be most careful about what type of scientific arguments we use for nature conservation politics, especially when talking about the design of nature reserves. It may be unwise to suggest any type of minimum area or minimum population-concepts at this stage of the discussion. The better approach might be to elaborate on the ethical and social awareness and choices; to reform the overall landscape management practices on a very broad basis in order to Integrate all possible measures needed to protect the biological diversity of our planet.

Zusammenfassung

Wachsende Bemühungen um einen wirkungsvollen Naturschutz können überall in der Welt mit der Geschwindigkeit des Ressourcenverbrauchs und der Naturzerstörung nicht mehr Schritt halten. Nur großflächige, von der Völkergemeinschaft gemeinsam getragene Schutzmaßnahmen vermögen möglicherweise diesen Prozess zu beenden.

Wesentlichen Anteil an der Naturverarmung haben die Isolations- und Zersplitterungsvorgänge, die auf Landschaften und auf Populationen von Tier- und Pflanzenarten einwirken. Die Europäische Gemeinschaft strebt eine Richtlinie zum Schutz der natürlichen Habitate und der wildlebenden Tier- und Pflanzenarten an und will dabei ein europäisches Verbund- system entwickeln, das den Namen NATURA 2000 tragen soll.

Die Richtlinie stellt hohe Anforderungen an die Mitgliedstaaten in Form von Flächenbeiträgen und betont die Schutznotwendigkeit der besonders seltenen und gefährdeten Lebensräume.

References

BORCHERT J.,(1980): Landwirtschaftliches Wegenetz und Gehölzbesatz in ausgewählten Gebieten der rheinischen Agrarlandschaft. Natur u.Landschaft 55 (10):380-385.

CURTIS, J.T.(1956): The modification of mid-latitude grasslands and forests by man. In: W.L.Thomas (ed.), Man's role in changing the face of the earth. Univ.of Chicago press.

DIAMOND, J.M.(1975): The island dilemma: lessons from modern biogeographic studies for the design of natural reserves. Biol.Cons.7:129-146.

DIAMOND, J.M., K.D. BISHOP & S.van BALEN (1987): Bird survival in an isolated Javan woodland: Island or mirror ? Conservation Biology 1(2):132-142.

EWALD, K.C. (1978): Der Landschaftswandel. Zur Veränderung schweizerischer Kulturlandschaften in 20. Jahrhundert.Ber.d.Eidg.Anstalt f.d.forstl.Versuchswesen 191.

FRANKEL, O.H. & M.E. SOULÉ (1981): Conservation and evolution.Cambridge University Press.

FRANKLIN, I.R. (1980): Evolutionary change in small populations. In: Soulé & Wilcox (eds.), Conservation Biology Sinaueer Ass.,Sunderland,Mass.

GRUMBINE, R.E. (1990): Viable populations, Reserve Size, and Federal Lands Management: A Critique.Conservation Biology, 4(2):127-134.

HARRIS,R.B.,MAGUIRE,L.A. & SHAFFER,M.L. (1987): Sample sizes for minimum viable population estimation.Conservation Biology, 1(1):72-76.

HEYDEMANN, B. (1981): Zur Frage der Flächengröße für den Arten- und Ökosystemschutz.Jb.Natursch.u.Landschaftspfl. 30:21-51.

JEDICKE, E. (1990): Biotopverbund. Grundlagen und Maßnahmen einer neuen Naturschutzstrategie. Ulmer, Stuttgart.

KNAUER, N.(1985): Bodennutzung, Landbewirtschaftung, Landschaftsgestaltung. VDLUFA-Schrr.,16:47-58.

LANDE, R. (1988): Genetics and Demography in Biological Conservation. Science 241: 1455-60.

LAHTI, T. & E. RANTA (1985): The SLOSS principle and conservation practice. Oikos 44: 369-370.

MACARTHUR, R.H. & E.O.WILSON (1963): An equilibrium theory of insular zoogoegraphy. Evolution 17: 373-387.

MCCOY, E.D. (1985): The application of island biogeographic theory to patches of habitat: how much land is enough ?. Biol.Conserv.25:53-61.

MADER, H.-J. (1980): Die Verinselung der Landschaft aus tierökologischer Sicht. Natur u. Landschaft 55(3):91-96.

MADER, H.-J. (1984): Animal habitat isolation by roads and agricultural fields. Biol. Conserv. 29:81-96.

MADER, H.-J. (1988): Effects of increased spatial heterogeneity on the biocenosis in rural landscapes. Ecol.Bull. 39:169-179.

MENGES, E.C. (1990): Population Viability Analysis for an Endangered Plant. Conservation Biology 4(1):52-61.

MURPHY, D.D. & B.A.WILCOX (1986): On island biogeography and conservation. Oikos 47:385-387.

MURPHY, D.D., K.E.FREAS, & S.B.WEISS (1990): An Environment Approach to Population Viability Analysis for a Threatened Invertebrate. Conservation Biology 4(1):41-51.

NOWAK, E. (1986): Internationaler Biotopverbund für wandernde Tierarten. Laufener Seminarbeiträge (ANL) 10/86:116-128.

QUINN, J.F. & S.P. HARRISON (1988): Effects of habitat fragmentation and isolation on species richness: evidence from biogeographic patterns. Oecologia (Berl.) 75:132-140.

SHAFFER, M.L. (1981): Minimum population sizes for species conservation. BioScience 31:131-134.

SIMBERLOFF, D. & L.G. ABELE (1976): Island biogeography theory and conservation practice. Science 191:285-286.

SIMBERLOFF, D. (1986): Design of nature reserves. In: Wildlifeconservation evaluation, (M.B.USHER ed.),Chapman & Hall,London.

SOULÉ, M.E. (1987): Viable populations for conservation. Cambridge university press, Cambridge.

Species Conservation: A Population-Biological Approach
A. SEITZ & V. LOESCHCKE (eds.) © 1991 Birkhäuser Verlag, Basel

Epilogue

V. Loeschcke[(1)] **and J. Tomiuk**[(1)(2)]
[(1)] Institute of Ecology and Genetics, University of Aarhus, DK-8000 Aarhus C, Denmark,
[(2)] Abteilung für Klinische Genetik, Universität Tübingen, D-7400 Tübingen 1, Germany

The widespread occurrence of species extinction is of major concern among biologists and, naturally, even more so among conservation biologists. This concern has led to different scientific activities and to the publication of several recent volumes on this topic (e.g. those edited by SOULÉ & WILCOX 1980; FRANKEL & SOULÉ 1981; SCHONEWALD-COX et al. 1983; SOULÉ 1986, 1987; WILSON 1988; SOULÉ & KOHM 1989).

The present volume differs from others in that most of the contributions deal with examples or issues directly relevant to the conservation of endangered species in Germany, as well as the rest of Europe. A few non-Europeans have joined those addressing the problems of general concern for conservation biology. Extinction is a global process with at least partly global causes and implications, so international cooperation is of great importance. However, some differences in the type of ecosystem, in scale, and in political history exist between the situation in Europe and in e.g. America, Southeast Asia, or Australia (see STEARNS et al. 1990 for a comparative evaluation of Swiss and American concerns). In central Europe no more large, coherent natural environments exist and the existing semi-natural habitats consist of limited sized patches separated by man-dominated areas. Though loss of species in absolute numbers is much higher in the tropics, the red lists on endangered species in e.g. Germany demonstrate that a high proportion of the existing species in Europe is also threatened.

The contributions to this volume are all written from a population biological viewpoint. Slightly different opinions exist as to which unit should be preserved in order to maintain populations of particular species. Though in principle it is generally accepted that the best unit of conservation is the whole ecosystem, most of the contributors to this volume, albeit mainly for practical purposes or for the sake of clarity, would prefer to concentrate on target species and corresponding habitat requirements, as well as the subsystem that encloses those species that have direct interactions with the target species. But how can we identify the subsystem which allows persistence of the species in question? What actually is the time scale of concern and what risk of extinction are we willing to tolerate? Over geological time all species are doomed to die out. No species can persist forever. The time scale in conservation biology, however, is a few thousand years at most. For that time period or even shorter ones as e.g. 100 years, we have to aim for a probabilistic description of the risk of extinction based on observations of the distribution of demographic, genetic and behavioral parameters (AKÇAKAYA & GINZBURG, this vol.). Unfortunately, in many instances neither the time nor the funds needed to aquire detailed knowledge of these parameters are available. Still, even on the basis of rather limited information, it is sometimes possible to establish reasonable first estimates of space requirements, population trends and probable effects of habitat fragmentation (Schonewald-Cox and Buechner, this vol.), which may be used as guidelines before further information can be assembled (see also MÜHLENBERG *et al.*, this vol.).

As the survival of certain species is often closely linked to the presence or absence of particular other species (VOGEL & WESTERKAMP, this vol.), or to the species composition in the biological community (LOESCHCKE 1987), the relevant biotic interactions of a target species with other species in the community should if possible also be revealed. In plant-pollinator interactions the occurrence of the host is often of vital importance to their pollinators. Furthermore, the persistence of many insect species is dependent on the existence of at least one particular plant family or even a single genus (VOGEL & WESTERKAMP, this vol.). Other species are often dependent on the persistence of particular habitat spots, as e.g. birds that show high degrees of philopatry and that breed in natal or adjacent nests (RISTOW *et al.*, this vol.), or that need resting sites during their seasonal migration.

Another question tackled in this volume is the role of genetics in conservation issues. Do species become extinct due to a too low level of genetic variation, and if so, how much variation is necessary to maintain the potential to persist? What type of population subdivision is most appropriate for the preservation of genetic variability? Does this subdivision coincide with the one suggested from an ecological point of view, and are not demographic and other ecological factors far more important for a species' survival ability, genetics being only a tool used to identify the demographic unit of concern?

Classically, genetics in conservation biology has focussed on how to best preserve genetic variability within a threatened species. The underlying assumption is that the amount of genetic variation is positively correlated with population size (BIJLSMA *et al.*, this vol.) and that higher genetic variability enhances adaptive flexibility and thereby the probability of survival of populations in the long run (AVISE 1989). Most conservation biologists, however, think that extinction is fundamentally a demographic process, at least in the short run (LANDE 1988), though it is granted that this process is influenced by genetic and environmental factors. But the close connection and mutual dependence between genetic and demographic factors has often been neglected and constitute one of the important areas of future research in conservation biology (LANDE 1988; EWENS 1990). The interaction between demographic and genetic factors may thus further diminish survival probabilities as compared with the pure demographic effects (GABRIEL *et al.*, this vol.). But the role of genetics is more complex. Genetic techniques may play an important role in defining the demographic or breeding unit   for the definition of management units (MENZIES, this vol.; TEMPLETON, this vol.; AVISE 1989). They may also be used to detect and monitor hybridization, the incidence of which has been increased by human activities through the introduction of "exotic" individuals or disturbance of isolating barriers (TEMPLETON, this vol.). Finally, the classical role of genetics in the management of small semi-natural or captive animal populations of endangered species is still to minimize inbreeding in the short run and to maintain adaptive flexibility in the long run (TEMPLETON, this vol.). The preservation of adaptive flexibility is also the main objective of gene conservation in cultivated plants and their wild relatives (HENRY et al., this vol.; GREGORIUS, this vol.). To prevent genetic variability from being "frozen" for too long and becoming maladapted to future environmental conditions, a dynamic conservation practice is preferable to static conservation in seed banks, though in practice a mixture of strategies is often more realistic (HENRY et al., this vol.).

In the context of preserving genetic adaptability, critical minimum effective population sizes have sometimes been suggested in the literature and have been applied to conservation programs. These numbers are only based on a number of genetic assumptions (for a critical overview and discussion see LANDE & BARROWCLOUGH 1987 and LANDE 1988) and do not incorporate any demographic processes (but see EWENS 1990) of major importance, especially in small natural or semi-natural populations. Here a population viability analysis (MÜHLENBERG et al., this vol.) or an ecological risk analysis (AKÇAKAYA & GINZBURG, this vol.) are generally to be preferred, as they take important population specific parameters into account (see also EWENS 1990).

Other important population specific parameters, besides the (inbreeding or mutation) effective population size, and the intrinsic growth rate and its variance, are the population subdivision and the dispersal rate among subpopulations or patches (AKÇAKAYA & GINZBURG, this vol.; HENRY et al., this vol; Mader, this vol.; OLDHAM & SWAN, this vol.; RENNAU, this vol.; RISTOW et al., this vol.). Natural populations are mostly non-homogeneous entities, and consist of a number of local subentities that interchange individuals. The proper system to consider then is the system of local populations, the metapopulation. Stochastic events will still effect local populations, though the result on the metapopulation level is of major importance: extinction of the whole system or not.

A positive correlation between area size and species number (or number of genera and families) under undisturbed environmental conditions has been established. The area of natural or semi-natural habitats is limited, and habitat fragmentation and destruction inevitably continue as a consequence of the increase in human population and its activities. A reduction in area size leads to a reduction in species number and to a lower average population size in the remaining populations. The dynamics of the extinction process following an area reduction are not yet fully understood, but the consequent loss of diversity is usually a slow process, the inital cause of which is not easily recognized. In this case, the link between cause and effect is not always immediately obvious. Even if the fragmentation and destruction of habitats were to be stopped at once, the decrease in species diversity would continue at least for some decades.

No general solution exists to the classic controversy on whether a single large or several small reserves of the same total size are better suited to preserve species from extinction (MÜHLENBERG et al., this vol.; AKÇAKAYA & GINZBURG, this vol.; HANSKI 1989), even if the choice between these alternatives normally does not exist in practice. The controversy, however, has lost some of its actuality, as the issue of conservation today is no longer a preservation of just a certain number of species per se, but rather the preservation of a particular species and corresponding suitable habitats. Nevertheless, it will often be useful to aim for several essentially independently viable populations of the target species, i.e. those in a number of suitable but geographically separated habitats. This strategy allows for recolonization if single populations become extinct, and provides some protection against catastrophes or harmful diseases that could cause extinction of one single large population.

A controversial issue in conservation today is whether it is worthwhile to concentrate one´s efforts on species, whose natural environment has disappeared; such efforts could prove to be futile, if the necessary habitat requirements of the species are not restored quickly. However, in cases of extreme longevity of the target species, the issue might be reconsidered if some hope for its successful preservation remains (BAUER, this vol.).

Man-induced changes in fauna and vegetation do not necessarily only mean extinction of species through fragmentation and destruction of habitats, but could also lead to the addition of new species through importation and thereby to an increase in species diversity (STARFINGER, this vol.). Such new introduced species might then become a threat to the native fauna and vegetation, and nature conservation then also has to focus on "dangerous" species that are able to invade and spread successfully (STARFINGER, this vol.). At present, however, we cannot yet predict beforehand whether an introduced species will be able to establish itself and eventually even become a pest. This knowledge could also be of importance in the evaluation of the chance of persistence of species after reintroduction.

Finally, conservation is as much a political issue as a biological one. Biologists can help in protecting species from extinction, but the long-term preservation of biological diversity depends on political decisions, legislation, available funding, the public attitude towards nature conservation (OLDHAM & SWAN, this vol.), and last but not least, on whether the human population will continue to grow in size and demands.

Acknowledgements

We are grateful to Prof. Stuart Barker for comments on the manuscript. Part of the final version was written while V.L. was a Visiting Research Fellow with Prof. Barker at the Department of Animal Science, Univ. of New England, Armidale, Australia. He wishes to express his gratitude for the hospitality received. The stay at Armidale was made possible by a grant from the Danish Research Council (Dansk Forskningsråd).

References:

AVISE, J.C. 1989. A role of molecular genetics in the recognition and conservation of endangered species. Trends in Ecol. and Evol. 4, 279-281.

EWENS, W.J. 1990. the minimum viable population size as a genetic and a demographic unit, pp. 307-316 in J. ADAMS, D.A. LAM, A.I. HERMALIN, & P.E. SMOUSE (eds.) Convergent issues in genetics and demography. Oxford Univ. Press, N.Y.

FRANKEL, O.H., & M.E. SOULÉ, (eds.) 1981. Conservation and evolution. Cambridge Univ. Press, N.Y.

HANSKI, I. 1989. Metapopulation dynamics: does it help to have more of the same. Trends in Ecol. and Evol. 4, 113-114.

LANDE, R. 1988. Genetics and demography in biological conservation. Science 241, 1455-1460.

LANDE, R. and G.F. BARROWCLOUGH 1977. Effective population size, genetic variation, and their use in population management, pp. 87-123 in M.E. SOULÉ (ed.), Viable populations for conservation. Cambridge Univ. Press, N.Y.

LOESCHCKE, V 1987. Niche structure and evolution in ecosystems, pp. 320-332 in E.D. SCHULZE & H. ZWÖLFER (eds.), Potentials and limitations of ecosystem research. Springer, Berlin-Heidelberg-New York.

SCHONEWALD-COX, C.M., S.M. CHAMBERS, B. MACBRYDE, & L. THOMAS (eds.) 1983. Genetics and conservation: a reference for managing wild animal and plant populations. Benjamin-Cummings, London.

SOULÉ, M.E. (ed.) 1986. Conservation biology, the science of scarcity and diversity. Sinauer, Sunderland, Mass. SOULÉ, M.E. (ed.) 1987. Viable populations for conservation. Cambridge Univ. Press, N.Y.

SOULÉ, M.E., & K.A. KOHM (eds.) 1989. Research priorities for conservation biology. Island Press, Washington D.C.

SOULÉ, M.E., & B.A. WILCOX (eds.) 1980. Conservation biology, an evolutionary-ecological perspective. Sinauer, Sunderland, MA.

STEARNS, S.C., B. BAUR, A.J. VAN NOORDWIJK, and P. SCHMID-HEMPEL 1990. Report to the Swiss Science Council on biodiversity: a critical review of the U.S. Office of Technology Assessment (OTA) report with recommendations for research funded by Switzerland, pp. 46-74 in   Technologien   zur   Erhaltung   der   biologischen   Vielfalt,   Schweizerischer Wissenschaftsrat (ed.), Bern.

WILSON, E.O. (ed.) 1988. Biodiversity. National Academy Press, Washington, D.C.